D1346922

The 'Atlante Farnesiano', carved about 300 B.C.

PLANETS, STARS AND GALAXIES

Descriptive Astronomy for Beginners

by

COMMANDER A. E. FANNING

M.B.E., D.S.C., F. R.A.S., R.N.

American Edition Revised by

DONALD H. MENZEL

Harvard College Observatory

DOVER PUBLICATIONS, INC.
NEW YORK

This Dover edition, first published in 1966, is a revised version of the work originally published by Mac Gibbon and Kee in 1963 under the title *Astronomy Explained*. It is published by special arrangement with Mac Gibbon and Kee (Arco Publications, Ltd.), 9 Grape Street, London, England.

This edition also contains a new Introduction by Donald H. Menzel.

Library of Congress Catalog Card Number: 66-24131

Manufactured in the United States of America
Dover Publications, Inc.
180 Varick Street
New York, N. Y. 10014

INTRODUCTION TO THE DOVER EDITION

THE past decade has witnessed an explosive growth in man's knowledge of the universe. New techniques, special lenses of original design, larger telescopes, radio and radar observations, electronic detectors and amplifiers, high-speed computers, satellites and space probes—all these and more have contributed to our increased understanding of the world we live in. The fact is that new books on astronomy have tended to be out-of-date even before they are published.

No one can slow down the advance of knowledge. However, the author and I have made every effort to include the latest information about the universe. The exploding sun takes on new significance in terms of the violent solar wind, which blows in gusts throughout the solar system. Interplanetary radar enormously increases the accuracy of measurements of distances within the solar system. Radio waves escaping from the heavy clouds of Venus indicate the existence of an extremely hot solid surface. Moon probes have photographed the far side of the moon and fine details of the surface facing the earth. The recent probe to Mars revealed a surface cratered like that of the moon. The Quasars, distant powerful sources of radio emission, have confounded the theoretical astronomer with problems of their nature and even of the origin and evolution of the universe.

The acquisition of astronomical knowledge still runs apace. But editorial changes made even in the galley proofs of this book will assure the reader that the content is as up-to-date as possible. I have been particularly impressed with Commander Fanning's ability to explain astronomy, simply, clearly, accurately. I think you will enjoy this book for an authoritative presentation of the very latest in astronomical science.

DONALD H. MENZEL

Cambridge, Massachusetts
January, 1966

5

PREFACE

—••€)€3••—

NATURE has few sights which can compare with the radiant beauty of a clear night sky. Yet today, how few ever savour that beauty. This book has been written mainly for those without deep astronomical knowledge in the hope that through greater understanding they may more fully enjoy the splendour of the heavens.

Astronomy is a fascinating pastime for young and old alike, and one which it is never too late to start. My aim has been to give, in this simple narrative, a comprehensive picture of the present extent of our astronomical knowledge coupled with a brief insight into some of the methods used by modern astronomers. For the benefit of the general reader it has been kept as 'non technical' as possible. Nevertheless it is hoped that it will provide a firm foundation on which those who are sufficiently interested may build, should the joys of discovering astronomy lead them on to wider study and further exploration.

My gratitude is due to Mr K. H. Fea of University College, London, who has so painstakingly checked the text. Also to the Directors of the London Planetarium who, for the past three years, have allowed me the pleasure of passing on the wonders of astronomy to the general public. It was this gratifying experience which led me to write this book.

CONTENTS

—••E)(3••—

9

LIST OF PLATES

The 'Atlante Farnesiano'   *Frontispiece*

Between pages 96-97

ILLUSTRATIONS IN THE TEXT

—··∙ℰ)(ℨ∙··—

CHAPTER I

—••£)(3••—

FROM THE BEGINNING OF TIME

These diamond orbs their various circles trace,
And run incessantly their daily race.
Round a fix'd axis roll the starry skies:
Earth, even balanced in the centre lies.
One pole far south is hid from mortal eye,
One o'er our northern ocean rises high.

THE ORIGINS of the study of astronomy are lost in the mists of time. When primitive man first attempted to arrange the pattern of his life his attention was naturally drawn to the heavens for therein lay the key to all order: the regular succession of day and night and the steady recurrence of the seasons. We know that he understood little of the whys and wherefores of it all; indeed the daily struggle for existence left him little time to ponder. It is small wonder that he readily accepted the sun, moon, stars and planets as deities to be worshipped for the benefits they might bestow, but at the same time to be feared and placated, lest they unleash the full furies of nature upon him.

To-day, as we are swept relentlessly forward into the space age, we smile at these primitive notions. From our earliest childhood we have learnt that our earth is a planet, rotating on its axis thereby giving us day and night: that its revolution around the sun once in a year gives us the seasons. We are so certain that these things are true that it is easy to forget that such beliefs were heresy only three and a half centuries ago.

Similarly, it would be inconceivable for our modern, highly complex twentieth century civilisation to exist without an accurate and comprehensive time system and a well established calendar, yet we would do well to remember that these conveniences were unknown in Elizabethan times. For centuries men had relied on burning candles, sand-glasses or, at best, weight-driven clocks, whose very nature precluded extreme accuracy. The fact that the period of swing of a pendulum is constant and depends on its length was demonstrated by Galileo in 1581, but the principle was not used to regulate a clock until nearly a century later when a Dutch astronomer named Huygens developed the 'escapement'. Only then did reliable methods of time-keeping become possible. Our present form of calendar, known as the Gregorian calendar, came even later, and was only adopted in England in 1752.

An amusing story is told of its introduction. The Julian Calendar, which had been in force since A.D. 325, incorporated a slight inaccuracy which, by 1752, had resulted in it being eleven days 'wrong'. When the new calendar was introduced, this was corrected by the simple expedient of missing out eleven days, and 2nd September 1752 was followed by 14th September. The peasant folk of the period obviously understood little of what was going on, but thought they were being cheated of part of their lives. It is reported that riots immediately broke out throughout the country, with the slogan 'Give us back our eleven days'!

Improvements in our time reckoning have come hand in hand with the advance of astronomical knowledge, for although the early study of astronomy can be traced back through thousands of years to the Chaldean astrologers of ancient Babylon and to the early Chinese, it is only in the last 350 years that any significant development has occurred. The fateful day in 1609, when Galileo Galilei, Professor of Mathematics at the University of Padua, first turned his telescope towards the skies, marked the end of an era almost as old as man himself. It marked, too, the beginning of an age of astronomical discovery that ever since has swept forward like a tidal-wave. Progress to-day is so rapid, and on such a wide front, that it is doubtful if even the astronomers who are themselves leading the advance can truly follow all its facets. The rest of us must be content to try and get the broad picture, filling in a little detail here and there as our knowledge improves. In this way we can at least begin to appreciate the magnitude of what has been achieved and possibly attempt to understand the significance of each new discovery that is made.

*   *   *

Let us take as our starting point the lines from *The Phaenomena* of Aratus, written more than 2,000 years ago, with which this chapter opens. These words surely express the sentiments which strike us first, even to-day, when we consider the heavens. This was how the ancients believed things to be. ' Round a fixed axis roll the starry skies ', and so indeed it appears. No wonder they thought the earth was the centre of creation; immobile, immovable, the pivot point of the whole universe.

This was the philosophy of the great Aristotle, certainly the most influential, if unfortunately not the wisest, of the early Greek philosophers. It is a measure of the stature of this amazing man that his astronomical doctrines, enunciated in the fourth century B.C., remained virtually unopposed for two thousand years. It is true that there were occasional men of vision who thought differently. A century before Aristotle the learned Pythagoras had taught that the earth was a rotating sphere. This view was later supported by Aristarchus of Samos, about 315 B.C., who also firmly believed that it revolved round the sun. But these men were far ahead of their time. Their teachings ran so contrary to man's natural instincts that they fell mainly on deaf ears. Such was the out-working of prejudice that after Aristarchus the anthropo-centric universe of Aristotle was not again seriously challenged for almost two millenniums.

It is interesting to note at this stage that because the earth was surrounded by the apparently rotating vault of heaven, the Greeks were led to portray the sky on globes centuries before anyone considered making a globe to represent the earth. One of the earliest of these ' star globes ' still in existence is the ' Atlante Farnesiano '[1] believed to have been carved about 300 B.C. This wonderful marble sphere some twenty-six inches in

[1] See Frontispiece.

diameter is borne on the shoulders of a kneeling Atlas, and may be seen in the National Museum in Naples. Although it shows very few actual stars, it is richly emblazoned with the forty-two mythological figures of the constellations mentioned in the *Phaenomena*. In the top left-hand corner of our picture can be seen the hero Perseus who, armed with the terrible head of Medusa, is in the act of rescuing the lovely Andromeda from the sea monster; Orion the Hunter can be seen waging his age-old battle with Taurus, the Bull, while three other constellations of the Zodiac, the Fishes, the Ram and the Heavenly Twins can also be plainly identified. The zodiac itself was always endowed with mystery and given great prominence, although it was naturally impossible to portray the planets, those ' wandering stars '[1] which moved within the zodiac, on a globe made of marble!

GREEK ASTRONOMY AND AFTER

Quite apart from their interest in mythology the ancient Greeks were undoubtedly the first people to attempt to treat astronomy as a true science. As early as the sixth century B.C., Thales of Miletus had suggested that the movements of the heavens might possibly be governed by perfectly natural causes, and should not be attributed solely to the whims and caprices of various Gods and Goddesses. Other great thinkers took up the idea, amongst them Anaxagoras, who explained the moon's phases and the reasons for eclipses; Pythagoras and Aristarchus, whose far-sighted teachings have already been mentioned; Eudoxus, who devised a complex system of rotating concentric spheres to re-present the movements of the heavens; Eratosthenes, who first measured the size of the earth; and finally the great Hipparchus, who has often been called ' the father of astronomy '. Hipparchus made wise use of the material gathered by great men before him, and about 120 B.C. produced the first star catalogue, containing about 1,080 stars, graded according to magnitude.[2] He also suggested a system of latitude and longitude for desig-

[1] *Planetes* = wanderer (Greek).
[2] See Page 131.

nating position on earth and is said to have discovered the phenomenon of the precession of the equinoxes.[1]

The last of the great Greeks of this era was Claudius Ptolemaeus, better known to us as Ptolemy, who worked at the famous observatory in Alexandria during the second century A.D. Ptolemy believed whole-heartedly in the doctrine of Aristotle. He was a brilliant mathematician and devoted much of his life to devising a complex but highly ingenious system, involving cycles and epicycles, which managed successfully to explain the strange and apparently erratic behaviour of the planets as observed from a supposedly stationary earth. His magnum opus, the *Almagest*, was a complete text book of astronomy, which remained a standard work for thirteen centuries. Although there were many things which the Ptolemaic system could not explain, it is doubtful if during the dark ages which followed many people ever troubled to understand it. Certainly no one seems to have questioned it. As its principles were later accepted as dogma by a none too liberal-minded church, this was perhaps just as well.

After Ptolemy's death astronomy seems to have remained static for more than 1,000 years. These were indeed the dark ages. Only for a brief period between seventh and tenth centuries A.D. do we find a temporary resurgence of interest in astronomical matters, and that from a most unexpected quarter. Early in this period the Mohammedan fanatics were ravaging their way through more than half the countries bordering on the Mediterranean, spreading death and destruction as they went. However, they seem to have been much impressed by the relics of Greek astronomy which they discovered and, in strange contrast to their trail of devastation, we find that many observatories sprang up in their wake in places as far apart as Baghdad in Mesopotamia and Toledo in Spain. Unfortunately these centres of learning taught only the astronomy of the Ptolemaic system, so that although a great deal of mathematical and scientific work was carried out their memory is preserved to-day chiefly in the many Arabic star-names still in use; also in our word ' Almanac ' which came from the Arabic ' Almanaca ', the

[1] See Page 61.

name given to the very detailed and accurate star maps that they made.

THE TRUTH DAWNS

Apart from this brief revival, however, the civilised world had been virtually oblivous to astronomy. Its great awakening is often connected with the voyages of those intrepid explorers Bartholomew Diaz, Vasco da Gama, Christopher Columbus and many others who sailed further from the land than had ever been done before. Of necessity these men relied on the information of the astronomers for their navigation, but they soon found that they were unable to ascertain their position at sea using the astronomical tables of the time. The old order, which had been accepted without question for so long, was at last proving to be inadequate. At the same time inconsistencies had become apparent in the calendar, which was clearly in need of revision. Gradually men began to realise that they must begin to think for themselves once more.

The wonderful story of the renaissance of astronomy has been told a thousand times, and its leading characters have their place in history. The actions and discoveries of these men laid the firm foundations on which later astronomers were to build. It is a fascinating tale, in which brilliance and determination eventually wore down prejudice and dogma, and led at last to the break-through in astronomy which has brought us to where we stand to-day. A very brief account of this early work is given in the pages which follow, in order that the reader may be able to view the past in its true perspective, and in the light of our present knowledge.

Among the eminent astronomers who were worried by the obvious inaccuracies of both the calendar and the astronomical tables was a Polish canon named Copernicus, a brilliant mathematician who eagerly applied himself to the task of finding out what was wrong. It was this far-sighted churchman who resurrected the centuries-old teachings of Pythagoras and Aristarchus and once again showed the world that the movements of the heavens could be explained far more simply if the sun and not the earth was considered to be at the centre of the system. He set down his beliefs in a book entitled *De Revolutionibus Orbium Coelestium*. Yet seldom in history can such wisdom and prudence have been combined. Copernicus was fully aware that the reception of his work was likely to be far from favourable. Man would not take kindly to being deposed from his supreme position at the centre of creation. Accordingly, although he completed his treatise in the year 1529 he appears to have delayed its publication until 1543, the very year in which he died. He thus successfully avoided the persecution and frustration which were to befall his enlightened followers, many of whom suffered much for their beliefs.

All this, however, was to come later. Astronomy in the second half of the sixteenth century was dominated by a less controversial figure in Tycho Brahe who, under the patronage of the King of Denmark, established an observatory on the island of Hveen, near Copenhagen. Although born three years after the death of Copernicus, Tycho Brahe was unfortunately no disciple of the new system. Indeed he had his own theories on the subject. Nevertheless he was fully aware that nothing could be proved without evidence and as a man of science he devoted his life to obtaining the necessary observations. For twenty years he kept systematic records of all that occurred in the sky, using instruments of his own design which far surpassed in accuracy anything that had previously existed. He charted the positions of the stars and, against this background, recorded the ceaseless wanderings of the planets and the moon. When he died in 1601 his work was still unfinished, but he bequeathed his priceless set of observations to his assistant, a brilliant German mathematician named Johannes Kepler, who immediately set about the quite stupendous task of analysing them.

Kepler himself believed whole-heartedly in the sun-centred system of Copernicus, but as his task proceeded he constantly found that the observations did not quite fit in with what should have been expected. Copernicus had insisted that the planets moved in true circles for, as in the Universe of Aristotle, he believed that everything in the heavens must be perfect, and the circle was regarded as the 'perfect figure'! This was strange reasoning for so great a man. The difficulty was finally resolved in 1609 when Kepler showed that if the orbits of the planets were considered to be ellipses rather than circles, the observations would fit perfectly. He went on to deduce his three famous laws of planetary motion which still form the basis of the celestial movements to-day.[1] These will be discussed in more detail when we explore the solar system, in Chapter III.

Here was another triumph for the protagonists of the new idea, but as yet there was no positive proof as to which system was correct. Ptolemy's explanation, whatever its complexities, could be made to fit the observed celestial movements very nearly as well as could the system of Copernicus, as modified by Kepler. Although many enlightened thinkers appreciated the force of the Copernican theory, they were unable to produce the necessary evidence to justify what was still regarded as a heretical doctrine. Proof came at last in the year 1610, from the telescope of Galileo.

It is a popular belief that Galileo invented the telescope. This, however, is not so. Its inventor was a Dutch spectacle-maker named Lippershey who, in 1608, showed that with a certain combination of lenses objects could be made to appear very much closer than they really were. In 1609 Galileo heard of the discovery and, obtaining details from the inventor, constructed his own instrument, his 'optic tube' as he called it, which he turned onto the heavens. It was like opening the door to a secret garden that had been locked since the beginning of time.

The wonders of the sky that were revealed to Galileo three and a half centuries ago were those which still delight the amateur astronomer to-day. Although his first telescope had barely the power of the average opera glass, yet it sufficed to show him that the Milky Way, which had puzzled men throughout the ages, consisted quite simply of myriads of beautiful, twinkling stars, too faint to be resolved with the naked eye. It showed that the moon was, in reality, an independent world, covered with mountains and plains; that the pure serene disc of the sun frequently displayed unsightly spots; that the planet Jupiter had, revolving around him, four tiny moons, a planetary

[1] See Page 88.

system in miniature. Above all it showed the phases of the planet Venus.

It had long been suspected that the planets were not 'bright orbs' like the stars, but that they shone simply by reflecting the light of the sun, as does the moon. If this were the case Venus and Mercury, which are both nearer to the sun than is the earth, should exhibit phases. Further, it could be shown that with the Ptolemaic system it would never be possible to see more than half the illuminated surface of these planets at any one time, and thus the phases would vary from 'new' to 'half' and back again. On the other hand, if the Copernican system were true, the phases would follow the complete cycle from 'new', through the gibbous (humped-backed) phase to 'full', and back again, in exactly the same way as does the moon. We can imagine what the feelings of Galileo must have been on the first occasion that he observed the gibbous phase of Venus. He alone, in all the world, possessed the necessary evidence to answer the most controversial astronomical question. Well might he have echoed the cry of Archimedes, nearly two thousand years before, 'Eureka! I have found it'.

The story of Galileo's conflict with the church of Rome and with his own conscience, for he was not only a scientist but also a devout churchman, has its place among the epic struggles of history. Despite his attempt to win papal support for his teaching by dedicating his main work, the *Dialogue*, to his Holiness, he was at length forced by the Inquisition to renounce his 'heresies'. Although we are told that he was well treated in his old age, and permitted to end his days as a prisoner in his own home, his powerful voice was never again heard in the great controversy. The *Dialogue* was placed on the 'Prohibited List' of books, whence it was not removed for two hundred years.

Galileo died in 1642 at the age of seventy-seven, a broken man who had seen his life's work scorned. We can only hope that in his heart he knew that the seed of truth he had sown must eventually take root and prosper, despite the temporary drought of disbelief. Even more than Copernicus, Galileo had been responsible for the first real break-through in man's understanding of the universe. Nevertheless it is doubtful whether his contribution was as great as that of the man who was to follow him in the pages of history, the shy, retiring genius, Sir Isaac Newton.

SIR ISAAC NEWTON

Born in the self-same year as the death of Galileo, Newton had, by the time he was twenty, acquired a fundamental understanding of the workings of the heavens far in advance of the knowledge of his day. We have seen how Kepler, from an analysis of the observations of Tycho Brahe, had been able to demonstrate *how* the planets behave in their orbits. Newton, with his famous theory of gravitation, endeavoured to explain *why* they should behave in this way. It is common knowledge that objects fall towards the earth due to a force which we call the force of gravity. Newton, however, maintained that this force operated throughout the entire universe. His theory of gravitation can be simply stated as follows : —

> ' *Every particle* in the universe attracts *every other particle*, with a force which is directly proportional to the mass of the bodies concerned but falls off according to the *square* of the distance between them.'

We are told that it was the motion of the moon which first caught Newton's imagination. Why should the moon continuously circle the earth? Most people will remember from their schooldays that Newton's first law of motion states that a body will remain at rest, or will continue to move at a constant speed in a straight line, unless it is acted upon by an *outside* force. The moon quite obviously does not continue moving in a straight line otherwise it would fly off at a tangent into space, like the weight on the end of a piece of string that has been swung round and released. Clearly there must be a force acting upon it which always exactly balances this tendency. Newton called this force *gravity*. He saw that the same considerations applied to every object on the surface of the earth, which would otherwise be thrown off into space owing to the centrifugal force of the earth's rotation; they applied also to the planets in their orbits round the sun, and to-day they apply equally to every artificial earth satellite which circles our globe.

Newton went on to provide mathematical proof that if his theory of gravitation was true, the bodies in the solar system would, in fact, behave in exactly the way observation had shown that they did.

Throughout the history of science these important words observation, theory and proof are constantly occurring, and it would be as well for us to digress for a moment in order to ensure that we understand clearly what they imply. By *observation* we mean quite simply the recording of events as they occur, or as they *appear to occur*, and this does not in any way necessitate an understanding of the reasons for their occurrence. For example, the sun can be observed to rise in the east, to swing across the sky, and to set in the west. The recorded observations would be the same, whether we consider the earth as rotating on its axis or the sun as circling the earth. Observation simply provides the data from which theories can be deduced.

A *theory* is a statement which attempts to rationalise and explain the observations, so that it is possible to say 'If this theory is correct, events will always occur in such and such a manner.' It may be checked by further observations which may either confirm it or show that it needs modification. Alternatively it may be necessary to reject the theory altogether and try again! In this connection it is always more prudent to consider a theory as *confirmed* rather than *proved*. Final proof is often very difficult to obtain and in the realms of astronomy will, in many cases, be impossible. The best that we can hope for is that our theories will continue to be confirmed by the results of further observations.

It is a remarkable tribute to Newton that his theory of gravitation, as originally set out in his great work, the *Principia*, stood the test of time for over two hundred years. Not only did it satisfactorily explain all later observations but, as we shall see in Chapter III, its brilliant use by other astronomers led directly to the discovery of the two outermost planets of the solar system, Neptune and Pluto. During the present century, however, the wider understanding of the structure of the universe that has been made possible by modern astronomy has led to the partial supercession of Newton's theories by the various theories of relativity proposed by Albert Einstein and others. Nevertheless, gravitation is something with which we all consider ourselves familiar, and which we can conceive in terms of our everyday experience. For simplicity, therefore, we will assume throughout most of this book that we live in a Newtonian universe. Only in the last chapter will it be necessary slightly to revise our ideas.

Newton's *Principia* is perhaps the most remarkable work in the history of astronomy, possibly of all science. Not only does it contain a full treatise on dynamics and motion which leads step by step to a complete mathematical explanation of all the celestial movements, but it also contains detailed studies on such diverse subjects as hydrostatics, waves, tides, and even the fundamental laws of optics. In each of these separate fields Newton's observations represented entirely original work far in advance of the understanding of his day. He might truthfully be described not as a genius but as six geniuses rolled into one. Through his unaided efforts man's scientific thought advanced further during his lifetime than it had done since the beginning of time. Hitherto the laws governing the physical universe were in many quarters either not understood or simply taken for granted (which amounted to much the same thing). Very little was known about the reasons *why* things happened as they did. Newton's great contributions were admirably summed up by Pope in the words:—

> Nature and Nature's laws lay hid in night:
> God said, ' Let Newton be! ', and all was light.

We can be sure, however, that Newton never saw himself in this way. Had it not been for the patient persuasion of Sir Edmund Halley, who had fortunately recognised his genius and urged him to commit his thoughts to paper, it is doubtful if the *Principia* would ever have been written. Newton solved his problems solely for his own satisfaction. Once they were solved they held no further interest for him, and the idea of putting his work on paper for the benefit of his fellow scientists would never have occurred to him. He was self-sufficient, and his attitude to the outside world tended to be one of complete indifference. Luckily, however, he had taken a liking to Halley and initially it was mainly to oblige his friend that the volumes of the *Principia* were written. Later, in a second edition of his work, he was led to add something of his own philosophy of life which shows him to have been a deeply religious man who recognised, as has many a great scientist since his day, the presence of a true God. Newton believed that it was inconceivable that the complex structure of the astronomical universe was not the result of intelligent design. He wrote ' This most beautiful system of the sun, planets and comets could only proceed from the counsel and dominion of an intelligent and powerful Being '. Yet he was always conscious of the efforts of others and in later life summed up his own contributions to science in this truly humble phrase: — ' If it has been given to me to see further than most men, it is because I have stood on the shoulders of giants.'

THE ROYAL GREENWICH OBSERVATORY

The seventeenth century had opened with a world torn by uncertainty, in which the discoveries of Kepler and Galileo were derided or, at best, regarded with suspicion. By its close all this had been forgotten. Observatories were springing up in many parts of Europe, and the study of astronomy was at last proceeding unfettered by dogma and prejudice.

Amongst the earliest of the great observatories to be established during this period was that at Greenwich, although its international fame was not to come until much later,[1] and the reason for its foundation was, once again, to help the mariners. Almost two centuries had passed since the seamen of Prince Henry the Navigator had experienced their early difficulties in fixing their position by astronomical means, yet the problem

[1] The meridian of Greenwich was universally accepted as longitude 0° in 1884.

was still far from solved. Briefly, it amounted to this. Latitude can be obtained quite simply by observing the height of the sun or a star at its highest point, which will occur when it is on the north-south meridian. This has been known since early times and its accuracy depends largely on the accuracy of the astronomical tables available. On the other hand, because of the rotation of the earth, the determination of longitude presents an entirely different problem, which can only be solved if the time is accurately known. The pendulum, which had made accurate time-keeping possible ashore, was quite useless at sea because of the rolling of the ship, and no other form of clock was nearly accurate enough. It was necessary to turn to the heavens themselves for the answer, and this appeared to lie in a study of the motion of the moon amongst the stars as she made her monthly circuit of the earth. Here, in effect, was the hand of a great clock which all could see. Provided that the moon's track amongst the stars could be accurately forecast, a glance at her position should always produce the right time.

This idea was excellent in theory but unfortunately the lunar tables of the day were very inaccurate. Even the positions of the stars, against which the path of the moon was to be traced, were still based on the observations of Tycho Brahe, made a good century earlier, and without the aid of a telescope.

Drastic action was needed. Britain, during the seventeenth century, had become a great sea-power with interests in America and the West Indies, and the safety of her seamen was, in consequence, of genuine concern to her King, Charles II. When the far-reaching results of these inaccuracies were explained to him he determined that ' he must have the stars anew observed, for the use of his seamen '.

Accordingly in 1675 the Royal Greenwich Observatory came to be built. It could never be said to have been a very lavish affair. The money for its construction was provided by the sale of spoilt gunpowder, while much of the timber and bricks came from the destruction of an old fort at Tilbury, considered to be no longer required. The first Astronomer Royal, John Flamsteed, was paid only £100 a year. He had no staff except his wife, who acted as his assistant, and he was expected to

provide his own instruments! Nevertheless, this was a beginning. Hitherto astronomy in England had been the province of
gifted amateurs and a few learned professors, but at last the
state had recognised its importance. Control of the new observatory was initially exercised by the Board of Ordnance (presumably because of the spoilt gunpowder!), but it was later placed
under the direction of the Board of Admiralty, a much more
natural arrangement in view of its connection with seamen, and
so it remained until 1965.[1]

So great was the importance attached to the problem of finding longitude at sea that in 1714 the British Government offered
a prize of £20,000 for a satisfactory practical solution. Almost
fifty years later, James Harrison, the son of a Yorkshire car-
penter, finally perfected a timepiece that would operate on
shipboard, and which became the fore-runner of the modern
marine chronometer. However, he never received the prize
offered for such a device. His invention also meant that the
lunar tables were no longer needed for their original purpose
of helping to tell the time. Nevertheless, the programme of
charting the skies, which was begun by Flamsteed, was con-
tinued by his successors and has in fact been carried on in one
form or another ever since. It is a task that is now shared by
many of the great observatories of the world, for astronomy is
a truly international science. Its history over the last 300 years
is studded with great names from all countries: Sir Edmund
Halley and Sir William Herschel, Cassini and Schiaparelli, Hale,
Hubble and Shapley, Kirchhoff and Ambartsumyan, and a host of
others, each of whom has added his contribution to our knowledge. The discoveries of these brilliant men and their colleagues
have given us a remarkably clear picture of the universe
around us, which it is our intention to explore stage by stage in
the chapters which follow. In order, however, that we shall
avoid the pitfall of ' not seeing the wood for the trees ', it would
be as well if we were to start by forming in our minds a picture

[1] Until 1965 the Astronomer Royal still appeared in the Navy List and
money for the work of the observatory was voted under the Navy Estimates.
The observatory is now, however, administered by the Science Research
Council and has been transferred to Herstmonceux, in Sussex.

of the whole vast concept, reduced to simple terms. As we proceed to study each aspect of the subject we shall then be able to picture it in its true perspective. Let us endeavour to produce a thumb-nail sketch of the universe.

THE SOLAR SYSTEM

We shall start with the earth—no longer the supreme centrepiece of the universe but, in the words of one of our poets:—

> . . . *but a speck of dust,*
> *Floating in desolate skies.*

The planet on which we live, a globe about 8,000 miles in diameter, is one of nine major planets revolving around the sun. Its companion, the moon, has a diameter roughly a quarter of that of the earth, and the two bodies keep approximately a quarter of a million miles apart.

The sun we shall for the time being describe simply as a globe of gas approximately 864,000 miles across and 93 million miles from the earth.

Already we find ourselves in trouble. While our experience enables us to form a fairly clear picture of distances of a few thousand miles, those involving millions are obviously completely outside our comprehension. Let us therefore take for comparison a yardstick which we can all understand. Let us consider a jet aircraft flying at 1,000 miles per hour.

It is well known that if an aircraft could fly at 1,000 miles an hour right round the equator of the earth, after just over a day it would arrive back at its starting point. If it were possible for it to fly at this speed to the moon, the journey would take approximately ten days. But to make a similar journey to the sun would require 3,875 days, or a little more than ten and a half years. Yet our earth is one of the planets comparatively close to the sun. Pluto, the furthest member of this far-flung family, is, on the average, about forty times as far away, or 427 years away by our jet plane. Clearly if we are to form any under-

possibility that one star might collide with another seems negligible and the late Sir James Jeans estimated that it might possibly occur once in 10,000,000,000 years, perhaps once since the universe began. By comparison, the congestion in our own solar system appears relatively crowded.

Returning to our bun ten miles in diameter, we find when we look at it from the ' top ' that it has the shape of a giant pin-wheel which, as we might expect, appears to be in rotation. Our own sun is situated in one of the spiral arms of the catherine wheel, about two-thirds of the way out from the centre. It is, of course, travelling round with the galaxy, and is taking approximately 200 million years to complete one revolution. The stars nearer the central hub have least far to travel and seem to be moving fastest. They thus complete a revolution rather more quickly than our sun whilst, conversely, those near the peri-meter move more slowly and are taking rather longer.

In spite of the thousands of millions of stars which make up this galaxy we know that the very sharpest eye, on the clearest of nights, is unlikely to be able to see more than about four thousand individual members without the aid of a telescope. In our ten-mile bun these would all lie within about 500 yards of the sun. Until the invention of the telescope nothing whatso-ever was known of what might lie beyond, for the rest of the galaxy, and practically the whole of the rest of the universe which lies outside our bun had been hidden and unsuspected since life began.

THE UNIVERSE

What does lie beyond our bun? The answer, quite simply, is more and more buns, millions and millions of them, possibly even a hundred thousand million, and each is another galaxy. They differ considerably in size and shape. Some appear to be flattish discs like our own galaxy, others are more elliptical; some, indeed, seem to be merely collections of stars with very little apparent form at all. On the average their size is com-parable with the galaxy which we have pictured as our own, each containing about one hundred thousand million stars. Once again we find that they are separated from us and from one another by enormous distances, even on the scale of our model. One of the nearest, known as the Great Galaxy in Andromeda, can just be seen on a clear night without the aid of a telescope, and in our model would be a bun slightly larger than our own, situated about 200 miles away. As we move outwards in all directions we find more and more buns, isolated islands in the depths of space and each one of them representing another galaxy. We cannot yet tell how far they extend or whether, in fact, they go on for ever, but already the world's largest telescope, the great 200-inch reflector at Mount Palomar in America, has photographed galaxies which, in our model, would be buns more than 200,000 miles away. Consequently we find that even on this fantastically small scale, in which our own glorious sun is nothing but a speck one millionth of an inch across, we have already extended our model almost as far as the moon! We can be certain, however, that even now we have only represented a tiny fraction of the universe. How much lies beyond we do not know, and at present it seems possible that we may never know. Ever since the days of Galileo, every increase in the power of telescopes has shown more and more galaxies. Not only is it certain that we are, as yet, nowhere near the limit, but the radio astronomers believe that they have already been able to 'observe' galaxies several times further away than those revealed by the 200-inch telescope.

At this point, lest the reader should be forming some mental picture of keen-eyed astronomers peering anxiously into space in search of ever fainter and more distant galaxies, it should perhaps be emphasised that in practice very little of an astronomer's observing time is spent in actually *looking* at the sky. Nearly all his work is done with the aid of a camera or its modern counter-part, the photo-electric photometer, combined with his telescope and often with many other ingenious instruments as well. The minute quantities of light that reach the earth from the more remote galaxies can only be detected by a very delicate photo-electric cell, and must then be multiplied many hundreds of times before they can be recorded and studied. As more sensitive photographic emulsions and more powerful multipliers become available, the limits of our observable horizon will certainly be extended still further. However, the most significant advances must now be expected from the radio astronomers. This science is still very much in its infancy, and, despite the impressive contributions it has already made to our knowledge, its present status must still be likened to that of optical astronomy during the decades immediately following Galileo's 'optic tube'. When, in the not too distant future, man at last breaks free from his earth-bound environment, and can establish his optical and radio observatories in space or on the surface of the moon, free from the blanketing effect of the earth's atmosphere, who can tell what depths may then be plumbed?

Before allowing our imagination to run away with us, let us return once more to our model and summarise our progress. We started with the sun, represented by a minute speck one-millionth of an inch across. Revolving around this speck is the earth, a minor member of our solar system, the whole of which is contained within a sphere the size of a pin-point. The sun turns out to be a rather minor specimen amongst the approximately one hundred thousand million suns in our galaxy, which is, itself, represented in the model by a bun ten miles in diameter and about two miles thick. From the earth we can, with our telescopes, observe millions of other buns, varying considerably in both shape and size, scattered throughout space out to distances of 200,000 miles in every direction. This represents the present limit of what we shall call the *optically* observable universe. If we were to include the information obtained by the radio telescopes, the dimensions of the model would need to be multiplied three or four times.

This is how our universe, as far as we know it, would appear if it were reduced in size some 50,000 million million times. Such a view may enable the reader to form a mental picture of the whole vast cosmos in perspective, so that planets, stars and galaxies appear in their correct relationship to one another and to the whole. Before going on to look at these things in greater detail, perhaps we should now digress in order to establish the *real* distances with which we shall be dealing.

LIGHT-YEARS

The earth, we have seen, is about ninety-three million miles from the sun, while the distance to the nearest star is more than a quarter of a million times as great, or approximately twenty-five million million miles. Quite obviously a new measurement is going to be required if the distances to the more remote stars or between the galaxies are to make any sense, and for this purpose astronomers make use of the speed of light.

Light is the fastest thing we know, and the first reasonably successful attempt at determining its velocity was made by Roemer as long ago as 1675.[1] Since then many ingenious experiments have been performed with the same object, the present accepted value being about 186,282[2] miles per second, or roughly half a million times as fast as a rifle bullet. Quite a simple calculation shows that the light from our sun takes about eight minutes to reach the earth. Consequently if it were possible, through some extraordinary catastrophe, for the sun suddenly to stop shining, we on earth would be unaware of the fact until eight minutes later, because the light that had already left the sun would still keep coming to us. Using the speed of light as our ruler we can conveniently say that the distance from the sun to the earth is eight *light-minutes*.

When we use the same measuring rod on the stars we find that the distance to the nearest star becomes a little over four *light-years*.[3] This is a star called Alpha Centauri situated in the southern sky, not far from the Southern Cross itself. It cannot be seen from the United States, except for Florida and those few southerly areas below 30° north latitude. A light-year is, quite simply, the distance a ray of light travels in a year at a speed of 186,282 miles per second, and in round figures this works out at a little under six million million miles. No wonder astronomers prefer to use this very convenient term in preference to dealing in miles.

One of the most familiar stars in the sky is our well-known

[1] See Page 108.

[2] We normally refer to this simply as 186,000.

[3] A comparison between this and eight light-minutes gives us a further excellent illustration of the extreme emptiness of space.

North pole star, Polaris, which marks the pivot point of the northern sky. Here we have a star which is about a hundred times as far away as Alpha Centauri, so that its distance is rather more than 400 light-years. It will be evident that if the light by which we are seeing Polaris has taken 400 years to reach us, we must be seeing the star not as it is now, but *as it was* 400 years ago. In terms of American history, we are seeing it as it was at about the time Columbus discovered America.

In our study of astronomy we shall therefore always be looking at the past. Sometimes it will be the fairly recent past, sometimes the very remote past, and this will depend entirely on the *distance* the object is from us. When, in the final chapter, we come to discuss how the universe began, this ability to look into the past will be seen to play a very important part.

The furthest stars which we can see as individuals with the naked eye are likely to be of the order of two or three thousand light-years away, but even this is only a beginning. The diameter of our galaxy which, in our model, was ten miles across, is in reality about 80,000 light-years. In other words a ray of light travelling 186,000 miles in every second would take about 80,000 years to travel from one side to the other. Our sun, we find, is situated in the outer regions of the galaxy, roughly 27,000 light-years from the centre. The Great Galaxy in Andromeda, still one of the nearest star cities to our own, is so remote that we are seeing it as it was approximately two million years ago, before even primitive man had appeared on earth. Finally, the most distant galaxies that have so far been observed, those which we found at the limits of our model, are believed to be at distances beyond 2,000 million light-years. We observe them, therefore, as they were at a time when the earth itself was very young. If, as we now believe, the radio telescopes are receiving signals from galaxies at several times this distance, it can only mean that the messages from these remote regions of space started on their travels before there even was an earth, let alone a man, to come to.

This is the universe we are now about to explore. It is a fascinating, challenging universe, secretive and yet not unkindly to the would-be detective who attempts to probe its mysteries.

CHAPTER II

—••E-)(-3••—

THE EARTH AND ITS CONSTANT COMPANION

O! Swear not by the moon, the inconstant moon,
That monthly changes in her circled orb.

TO SHAKESPEARE'S Juliet the moon was inconstant because of her monthly changes of shape. In reality, however, she is the most constant companion of our earth, as the two circle the sun together. They are inseparable, and both merit our special attention.

To the first men on the moon our earth will present a truly magnificent sight, a brilliant bluish disc or crescent, depending on its phase, shining in a jet-black sky. The earth, like the moon and planets, shines by reflecting the rays of the sun, so that to an observer on the moon it will appear to go through a complete cycle of phases from new to full and back again in approximately 29½ days. Unlike the moon, however, the earth is turning rapidly on its axis, so that the visible surface features, the oceans and continents, will appear to be continually changing.

We all know how bright a full moon can appear in our sky and how beautifully she illuminates the landscape with her soft light, yet the moon is really a very feeble reflector of sunlight. We could liken her to a tarnished mirror, capable of reflecting only about seven per cent of the sunlight which falls upon her surface. By contrast the earth is an excellent reflector, owing mainly to its wide oceans, its extensive polar ice caps and its prolific cloud cover. It probably reflects back into space nearly half the energy it receives from the sun. This reflective property is called the *albedo*, so that we say the albedo of the moon is about 7 per cent, whilst that of the earth is approximately 50 per cent. The diameter of the earth is about four times that of the moon, so that its surface area is some sixteen times as great. It is small wonder that the planet earth will look such a splendid object from outer space, for a 'full earth' will reflect more than a hundred times as much sunlight as a full moon, and, when visible in the sky, will certainly be sufficient to light the lunar explorers on their way during their fortnight-long nights.

As the light of the sun passes through the earth's atmosphere it is scattered by the molecules of gas and particles of dust which it encounters. We all know how the short waves or ripples in a pond are easily dispersed by obstructions where the longer waves of the ocean would continue their progress undeterred. In just the same way, because of its shorter wave-length, the light at the blue end of the spectrum suffers more from this scattering effect of the atmosphere than does the light at the red end and this is what gives us our blue day-time sky. We can regard it as a dazzling curtain of light which prevents us from seeing the fainter light of the stars and planets whenever the sun is above the horizon. If we had no atmosphere there would be no such scattering of light and away from the dazzling sun the stars could still be seen, even in the daytime. Each would appear as a brilliant point of light shining out of the absolute darkness. They would not twinkle, for this, too, is a product of disturbances in the atmosphere which cause the rays of light to be diffracted as they pass through it. Instead the stars, the planets, and even the sun himself, would shine steadily with incredible brilliance out of a sky of almost perfect blackness.

This is the view of the heavens that space travellers have already enjoyed. Even when they land on the surface of the moon they will still be untrammelled by any interfering atmosphere and, when the sun is absent, their sky will be dominated by the exciting bluish planet earth, which we are now going to examine.

THE ATMOSPHERE

From the surface of the earth, we survey the sky from the bottom of a great ocean of air many miles in thickness, the atmosphere, which affects our lives in many different ways. Although we normally take it for granted, we are really very fortunate to possess the type of atmosphere we do for, as we shall see in the next chapter, we should be very poorly placed if we were to exchange it for the atmosphere of any of the other planets. Let us therefore take a brief look at this very important mantle. It is comparatively dense in its lower regions and, because of the 'weight' of air above us, there is a pressure of about fifteen pounds per square inch acting on every part of our bodies. This pressure is actually called *one atmosphere*. It is as if we were supporting a sheet of armour-plating roughly nine feet thick and the total thrust on the average human body amounts to several tons. The majority of the atmosphere is packed tightly around the earth, and more than nine tenths are found in the first five miles above its surface. As we ascend, the pressure falls off rapidly and the air becomes more rarefied. Three and a half miles up in the Andes in South America we find the highest permanent human habitations and here the pressure is only half an atmosphere. On the top of Mount Everest, five and a half times up, it is one third, and ten miles up only one-tenth of its value at sea level. After that the decline is very rapid and at a height of fifty miles the pressure is only one hundred-thousandth of an atmosphere, as perfect a vacuum as can ever be produced in our laboratories. Nevertheless it is impossible to say precisely where the earth's atmosphere ends. Evidence from the Aurora Borealis[1] shows clearly that traces still exist at heights of 6-700 miles, and this has been amply confirmed by experiments carried aloft in satellites which have detected traces at distances of several thousand miles from the surface. It is probable that there is no sharply defined boundary between the atmosphere of the earth and that of the sun, for astronomers now believe that the earth itself is immersed in the sun's outer atmosphere, which may well extend throughout the entire solar system.

The air we breathe is a complex mixture of gases made up of eleven main constituents of which only two, oxygen and water vapour, are essential for supporting life. The remainder, nitrogen (78 per cent), argon, carbon dioxide, hydrogen, neon, krypton, helium, ozone and xenon, give the necessary dilution to the oxygen, but otherwise hardly affect us at all. Roughly one-fifth

[1] The Northern Lights—see Page 46.

(21 per cent) of the atmosphere is oxygen, a far higher proportion than is found in the atmospheres of any of the other planets, and fortunately for us this is continually being replenished by the abundant plant life which breathes out oxygen whilst at the same time absorbing the surplus carbon dioxide exhaled by other living creatures. Water vapour, on the other hand, is a far more variable quantity. On damp humid days the air we breathe may contain as much as 5 per cent of this essential commodity, whereas out in the dry arid wastes of the desert there may never be more than a trace. It practically disappears altogether above five miles and calculation has shown that if all the water vapour in the atmosphere were wrung out and condensed it would still only cover the earth's surface to a depth of about two and a half centimetres. Nevertheless its presence is vital, as life in any form in which we can conceive it would be quite impossible without some form of moisture.

It is convenient to consider the atmosphere as consisting of a number of separate layers, although, of course, there is really no hard and fast division between one and the next. The dense layer of air near the earth's surface is called the *troposphere*, a turbulent region which gives us our changeable weather. It extends to a height of about a dozen miles over the equator, but only three or four miles over the poles. Above it is the *stratosphere*, and higher still the *ionosphere*, which extends from about fifty miles up to several hundred. Finally, in the region where the atmosphere has practically ceased to exist, there is the *exosphere*, which extends indefinitely upward into the *magnetosphere*.

Conditions in the stratosphere are generally far more constant than we find them lower down, although very high winds, called jet streams, do exist and may reach 200 or 300 miles per hour. Much study has recently been devoted to these jet streams, not only because of their effect on high-flying aircraft, but also because it is now definitely established that, in the long term, it is the conditions in this region which govern our weather at the surface. Within the stratosphere there is a thin layer of particular significance, known as the ozone layer. People often talk of going to the seaside to benefit from the ' ozone ', which, in their minds, generally means the sea-breezes. These, in fact, hardly contain any ozone at all! Ozone is a special form of oxygen, each molecule of which contains three oxygen atoms instead of the normal two we find in the air we breathe. In some way this gas has the peculiar property of being practically opaque to ultra-violet light, which is extremely harmful to living tissue. Thus the ozone layer, which is situated only some twenty miles up in the atmosphere, plays a very vital part in the maintenance of life on this planet. In the words of the psalmist : —

So that the sun shall not burn thee by day . . .

Although we tend to think of the sun as a dispenser of light and heat, we must not forget that he is really radiating energy over a vast range of frequencies. The visible light by which we see covers only a minute part of the whole electro-magnetic

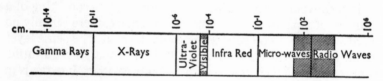

Fig. 1. The electro-magnetic spectrum, which runs from gamma rays to long radio waves. The wave-length is shown in centimetres. Only waves within the two shaded bands are able to penetrate the earth's atmosphere.

spectrum, which runs from the shortest gamma rays, through X-rays, ultra-violet, visible light, infra-red, and so into the wide range of radio waves. (See Figure 1.)

Because of its composition our atmosphere acts as a remarkably effective filter which stops nearly all waves except those in two narrow frequency bands from reaching the earth. We might regard it as having two small windows in it. One we have always known about as it lets through the light by which we see, and we might call it the optical window. The other, which we might call the radio window, only lets through radio waves covering a limited band of frequencies, and it was the discovery of this second window by an American named Janski in 1932 which eventually led to the development of the new science of radio astronomy.

Thus we see that far from being completely transparent, our atmosphere acts as an opaque shield, barring the way to almost all radiation from outside except in these two narrow frequency bands. The ozone layer effectively stops the very short wavelength radiation, while most longer wave-lengths are stopped much higher up, and are reflected back into space by the ionosphere.

The ionosphere is an extremely interesting region of the atmosphere, about which a great deal still remains to be learnt. It results from the action of the very short wave-length X-ray and ultra-violet radiation from the sun which causes some atoms to lose one or more of their electrons and so become *ionised*. A shower of such ionised atoms can act as a most effective screen to radio waves of certain frequencies, not only stopping them from reaching the earth from outer space but at the same time stopping similar waves from escaping.

The ionosphere thus acts like an enormous spherical mirror which completely surrounds the earth, and is polished both inside and out. Were it not for this mirror world-wide radio communication would be impossible, for radio waves, which travel in straight lines, would, on reaching the horizon, go straight on and disappear into space. On hitting the ionosphere, however, waves of certain frequencies are reflected back to earth again, and in this way it is possible to send messages right round the globe.

Unfortunately our spherical mirror is not completely permanent. At times when the sun is particularly active an excess of short-wave (X-ray and ultra-violet) radiation may temporarily shatter it, causing radio fade-outs, and impairing long-range communication over large areas. At these times the sun also sends out showers of very fast-moving electrically-charged particles, ions and electrons which, should they come in the direction of the earth, arrive about thirty hours later. They spiral around magnetic lines of force to the two magnetic poles. Radio communication is then further impaired by the violent electrical and magnetic disturbances which are caused and, as a spectacular by-product, we often get beautiful and colourful displays of the *Aurora Borealis* or *Aurora Australis*. As the electrified particles approach the earth they spiral in along the magnetic lines of force and bombard the thin upper air, causing it to glow like the gas in a 'neon' sign. Such phenomena are generally about a hundred and fifty miles up, but have been known to occur at heights of up to six or seven hundred miles.

As an extension to our communications systems there are now satellites circling the globe which act as relayers of messages, either actively or as reflectors. Communication across the Atlantic has also been achieved by bouncing signals off the moon. It goes without saying, however, that such systems will always be restricted to that band of radio frequencies which can penetrate our radio mirror, the ionosphere.

Before we leave this intriguing atmosphere there are two further ways in which it affects us which deserve special mention. We all know how hot we can get sitting in the sunshine behind a window-pane. The glass allows the radiation from the sun to pass through without itself becoming heated, while the radiation from inside, being at a lower temperature, is stopped by the glass. This is the reason why the temperature inside a greenhouse rises. The earth's atmosphere acts as the glass of a greenhouse, not only to give us an equitable temperature by day but to prevent it falling unduly at night through radiation. The highest temperature ever recorded on earth is 136°F.[1] while the lowest, — 109°F., was experienced by the Russian Antarctic expedition in 1958. These, however, are exceptional. Generally our temperature varies only within small limits which do not make it uncomfortable, but without our atmosphere we should suffer the same extremes of temperature as does the moon, ranging from about 220°F. (above the boiling point of water) at mid-day down to the equivalent of 220°F. below zero by night.

In another way, too, the atmosphere acts as a shield. If we look at the sky on any clear night our attention is certain at some time to be caught by a brilliant streak of light as a shooting star plunges to its glorious, fiery death. These streaks have, of course, nothing to do with the ordinary stars in the sky. The

[1] In N. Africa in 1922.

event we are witnessing is taking place in the ionosphere, per-haps 70 or 100 miles above our heads. A tiny meteor, probably only the size of a grain of sand, which for millions of years has been travelling in space as a minute planet of the sun, has suddenly collided with the earth. As it plunges into the rarefied upper atmosphere it compresses the air in front of it and so raises its temperature that the meteor quickly becomes white-hot, vaporises, and finally burns itself up in that glorious trail of light. Its remains drift gradually to earth in the form of a very fine dust. Occasionally a larger particle may plough its way into the denser layers of the atmosphere before burning up, and we then call it a *fire-ball*. Even larger ones may strike the earth's surface and are then called *meteorites*. Although we are nor-mally not aware of the process, several hundred tons of meteoric matter are deposited over the surface of the earth every year. Once again we see how fortunate we are to have a protective atmosphere to shield us from this constant bombardment.

We have devoted some time to examining the atmosphere because of the many ways in which it contributes to the life of mankind on this earth. In Chapter III we shall see how the type of atmosphere which we observe on the other planets can give us an immediate indication of the probability of life on the other worlds of our solar system. Let us therefore briefly sum-marise the benefits we enjoy as a result of this atmosphere. It provides the oxygen and water vapour we need to support life, while its blanketing effect ensures that we enjoy a reasonably equitable temperature. It protects us from the harmful ultra-violet rays, the cosmic rays and the intense bombardment of the millions of meteors and micro-meteorites whose traffic lanes we are constantly traversing. Finally, although this can hardly be described as essential to life, the ionosphere makes possible our world-wide radio network.

There are, of course, many other benefits we derive from the lucky chance of inheriting this atmosphere, but they do not concern us here. What does concern us, however, is the tremen-dous handicap, from the astronomer's point of view, of having to work through this great blanket of air which allows only a minute proportion of the energy reaching the earth from the stars to penetrate to our instruments. In the case of the hottest stars by far the greater part of the energy emitted lies in that very ultra-violet band which is so effectively blocked by the ozone layer. An astronomer's task can thus be likened to that of a musician, whose ear can only detect the notes in a single octave, trying to reconstruct a symphony whose notes cover the whole range of sonic frequencies.

With the exception of an occasional meteorite the earth itself is the only part of the universe which we can really study at first hand. It is the only laboratory specimen we can examine. The elements which we find on earth are those of which the stars are made, and it is a remarkable fact that every element identified elsewhere in the universe has also been found to exist in this tiny fragment. The laws of nature which are disclosed by our laboratory experiments must thus be taken to hold good throughout the universe, until we discover any evidence to the contrary.

INSIDE THE EARTH

Although much remains to be learnt about the atmosphere and, in particular, its upper layers in the ionosphere, our knowledge of these regions has increased immeasurably since 1957, when the first earth satellites began their exploration. By contrast our knowledge of what the inside of the earth is like must still be regarded as extremely sketchy. The world's deepest mines go down a mere 10,000 feet, or about two miles, and scarcely prick the skin. Even the deepest oil wells only penetrate to a depth of about four miles. Nevertheless a study of conditions down the mines shows at once that the temperature rises quite sharply the deeper we go. The average increase of temperature is about 87°F. for every mile. A descent of 1,000 feet, comparable to the world's tallest building, would raise the temperature by some 16°F. Down the world's deepest mines the walls cannot be touched for they are at a temperature above that of boiling water. If we assume that this increase continues, we must expect the centre of the earth to be very hot indeed, so hot in fact that it is bound to be in a molten state.

Although we cannot penetrate far into the earth, scientists have been able to learn a great deal about it by studying the behaviour of earthquake.waves. This is the science of *seismology*. The earth's crust is subject to continual stresses and strains which periodically cause it to fracture and the resulting shock waves can be detected by delicate instruments called seismographs, which measure and record the vibrations. From a study

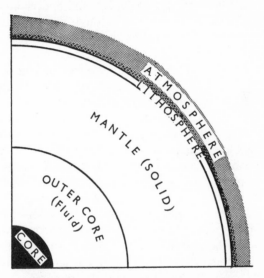

Fig. 2. Part of a section through the earth showing the various layers approximately to scale. The highest mountains and deepest seas would be less than the thickness of a pencil line. The dark tinted layer between the lithosphere and the atmosphere represents the hydrosphere.

of these records a fairly clear picture of the earth's interior has emerged, and this is believed to consist of a number of distinct layers, as is shown in Figure 2.

In the middle is a molten nickel-iron core approximately 1,550 miles in diameter, the centre of which is probably at a temperature of 7-8,000°C., rather hotter than the surface of the sun. Here the pressure is likely to be between three and four million atmospheres. This inner core is surrounded by an outer core about 1,380 miles thick consisting of metallic rocks, also in a fluid state, whilst above this comes a solid rocky mantle with a depth of some 1,800 miles. Finally there is the thin outer crust, the *lithosphere*, which varies in thickness from about thirty miles under the highest mountains to only two or three miles under the oceans. It is separated from the rocky mantle by what is called the *Mohorovic discontinuity*, and hence the ambitious American project to bore a hole through the crust in order to study the mantle has been called the *Mo-hole project*.

To complete the picture we should perhaps include the hydrosphere, the oceans and lakes which cover some four-fifths of the earth's crust, with a greatest depth of approximately seven miles and, enclosing it all, the atmosphere.

Before we leave the structure of the earth let us look once more at its magnetism. We all know that the earth behaves like a giant magnet and so enables us to use a magnetic compass to obtain direction. Unfortunately the magnetic poles are not quite at the true North and South poles of the earth and they are continually shifting their position over the years. Despite the fact that magnetic compasses have been in use for centuries and were certainly employed by the ancient Greeks as well as the early Chinese, the reasons for the magnetism of the earth are still not fully understood. It probably has its source in the fluid core which may operate like a gigantic dynamo. At the same time there are indications from the various magnetic rocks in the earth's crust that about a million years ago the North and South poles were the opposite way round to what they are today. Two million years before that, however, they were probably as we now know them, so that it seems possible that the steady change which we observe may be a stage in yet another reversal.

One of the most fascinating and far-reaching discoveries that have been made by the artificial earth satellites directly concerns the earth's magnetism. This is the existence of broad zones of intense radiation which surround the earth, and which are apparently caused by electrified particles trapped by its magnetic field. They have been named the Van Allen belts after

Professor James Van Allen of the United States of America, who was responsible for the instruments in the American satellite, Explorer I, which gave the first indication of their presence. Later satellites were used to trace out their extent and showed that the two main belts are situated, one about a thousand miles above the earth and the other, a much wider one, centred at about ten thousand miles. It is possible that there may be others. To enter these belts would be like moving into the close vicinity of an atom-bomb explosion where the radiation dosage might amount to more than 5,000 times what the human body could stand. Suitable protection or complete avoidance will thus become vital problems for the would-be space traveller.

The trapped particles in the Van Allen belts are believed to originate in the sun. Although the existence of these belts was not even suspected in the pre-satellite days, it seems likely their study will in the future provide much important evidence about the way in which the sun influences our planet moving within its atmosphere.

THE 'FIGURE' OF THE EARTH

The earth is frequently described as 'shaped like an orange' as it is slightly flattened at the poles, and while this general idea is correct, it should not be taken too literally! Certainly the earth's diameter measured from pole to pole is shorter than at the equator, but the difference is very small, amounting to only twenty-seven miles in about 8,000 (or one part in two hundred and ninety-seven). This is described as the earth's oblateness and is one result of its rotation on its axis. The precise values are 7,899·98 miles between the poles and 7,926·6 miles at the equator. Once again the artificial satellites have been able to add to our knowledge of the 'figure' of the earth by showing that it is really a slightly 'pear-shaped orange', bigger at one end than at the other. The difference is probably only a matter of about 150 feet, but is nevertheless significant. By the same analogy, the irregularities of the earth's surface are sometimes likened to those on the peel of the orange, but once again this is a gross exaggeration. If the earth's surface was as rough as the skin of an orange, the mountains would be fifty miles high. In fact the surface and figure of the earth far more nearly resemble those of a billiard ball than any orange.

The 'weight' of the earth is approximately six thousand million million million tons (6,000,000,000,000,000,000,000) and its average density is about five and a half times that of water. Again it is Newton's law of gravitation that makes these measurements possible. Newton said that every particle in the universe attracts every other particle; the earth attracts us or a football or any other object we care to choose, and we or the football attract the earth to *exactly the same extent*. The amount of attraction varies according to the combined masses of the two bodies concerned, but quite obviously while the attraction of the earth will have a marked effect on the football which we measure as the 'weight' of the football, the effect of the attraction of the football on the earth will be insignificant. If, however, a very massive body were placed near the football, the effect of its attraction on the football, although far less than that of the earth, might still be measurable. The earth was first 'weighed' in 1740 by a French astronomer named Bouguer, using this principle. He measured the amount by which a pendulum was deflected from the true vertical by the gravitational pull of a mountain and, by estimating the mass of the mountain, was able to express it as a fraction of the mass of the whole earth. The experiment has been repeated on many occasions since then, notably by the English Astronomer Royal, Maskelyne, in 1772, who made use of Mount Schiehallion in Scotland. More recently, however, very precise estimates of the mass of the earth have been obtained by laboratory methods, using large accurately known weights and extremely delicate balances and measuring equipment, but the principle used is exactly the same.

The mass of the earth can also be determined with the help of the moon. We have already seen that it is the gravitational pull between the earth and moon which prevents our satellite from flying off into space. Once the moon's distance has been measured[1] it is a comparatively simple matter to calculate the

[1] See Page 65.

pull required to restrain her, and hence the mass of the earth which will give this pull.

Fortunately the figure obtained agrees well with that determined from laboratory experiments so that we can, with confidence, extend this principle to 'weighing' the sun, the planets and even the stars. Just as the earth's pull on the moon can give us its mass, so the mass of the sun can be determined from the pull it exerts on the earth or, indeed, on any other planet. It is encouraging that all the planets give very much the same answer, and show that the sun has a mass approximately a third of a million times that of the earth.

THE DAYS AND THE YEARS

The earth has two main movements which affect our daily lives; its rotation on its axis and its annual journey round the sun, and although we are all well aware of these, we shall find that neither of them is quite as simple as it at first appears.

Let us first consider the earth's rotation which day by day causes the sun to appear to wheel over us from east to west. Instinctively we regard the sun as our timekeeper and when we say that our day lasts twenty-four hours, we really imply that the interval between his crossing our north-south meridian on successive days is twenty-four hours. During this interval, however, the earth has also moved on in its orbit around the sun. (See Figure 3.) In reality the earth completes one revolution *in space* in approximately 23 hours 56 minutes, and the remaining four minutes are taken in ' catching up the sun again '. If instead of considering the sun we were to time the interval between two successive crossings of a star over our meridian, this would be approximately 23 hours 56 minutes 4 seconds, which is called a *sidereal day*[1] and is very nearly constant. Only the advent of atomic clocks has shown that even the rate of spin of the earth fluctuates very slightly, and that the length of a sidereal day may sometimes vary by a few millionths of a second!

A far more obvious variation in the length of the day is that caused by the changing speed of the earth in its orbit. Kepler

[1] *Sidus* = star (Latin).

showed that the closer a planet is to the sun the faster it will move,[1] so that as the earth's distance from the sun varies between about 91½ and 94½ million miles, its speed changes appreciably. In consequence the amount by which the earth has to ' catch up the sun ' (Figure 3) varies continuously through the year, and this naturally affects the apparent length of the day.

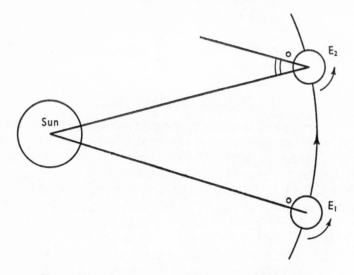

Fig. 3. Sidereal and solar days. When the earth is at E1, the sun is overhead at a point O. After completing one revolution *in space* the earth has reached E2, and a sidereal day has passed. Sun is not yet overhead at O, and earth must turn a little more before a solar day has passed.

Convenient as it might appear to regard the sun as our metronome, it would obviously be undesirable to have clocks which constantly altered their rate in order to keep pace with him. An average value for the length of the day has therefore been chosen which evens out the differences over the year. Only on certain dates does clock-time agree with sun-time. More normally the sun crosses the Greenwich meridian not *at* noon G.M.T. but a few minutes before or a few minutes after it. The

[1] See Page 88.

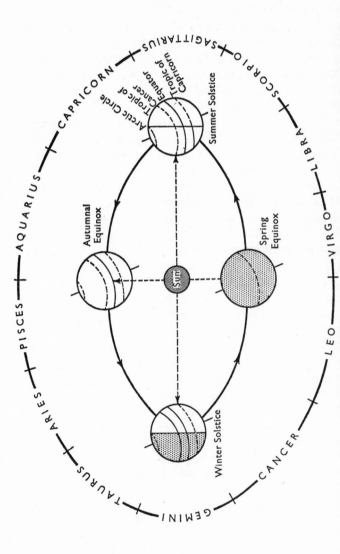

Fig. 4. Showing the earth's annual path round the sun, and the reason for the seasons, which have been labelled as they occur in the northern hemisphere. The earth's axis is inclined at approximately 23½° to the plane of its orbit. The twelve constellations of the zodiac are intended to represent the stars. By looking at the position of the earth at any season it can be seen which constellations are behind the sun, and therefore not visible, and which would be seen in the night sky. For example, at the time of the winter solstice, Capricornus, Sagittarius and Scorpio cannot be seen because they are in the same part of the sky as the sun, but Taurus, Gemini and Cancer are well seen.

discrepancy, which is known as the *equation of time*, has a maximum value of about 16 minutes and accounts for the fact that a sundial does not show the time accurately at all times of the year. As a timekeeper the sun is somewhat erratic!

Figure 4 shows, diagramatically, how the earth moves round the sun. The earth's axis is inclined at an angle of approximately 23½° to the plane of its orbit and all round, at great distances, are the stars.

At noon on midsummer's day the sun reaches his highest point for the year, which in New York is about 73° above the horizon. He can thus never be seen directly overhead in England, but on this day will appear in the zenith at noon in latitude 23½° North, which we call the *Tropic of Cancer*. (Figure 4.) Six months later, when it is mid-winter in the northern hemisphere he will be overhead at noon in latitude 23½° South, the *Tropic of Capricorn*, whilst in New York he does not rise more than 26° above the horizon on this day. Only at the equinoxes in the spring and the autumn can the sun appear exactly overhead on the equator itself.

While we take it for granted that it will be hotter in summer than in winter it is as well to be certain that we understand the reason for this. The higher the sun is in the sky the greater will be the proportion of his radiation which falls onto any given area of the earth's surface, as can be shown by a very simple experiment. If we hold a sheet of card one foot square in front of a bright lamp so that the rays fall directly upon it (i.e. to a fly on the card the lamp would be 'overhead') the card will intercept the rays of the lamp over a full square foot. If we now tilt the card away from the lamp, the amount of lamplight it will intercept will obviously decrease (and the shadow of the card will get smaller). The area of the card itself is still the same, but as less lamplight is now falling on it we can see that the amount of light striking each square inch must have been reduced. When we have tilted the card through 60° only *half* the initial amount of light will be falling onto it, and this is equivalent to considering a place 60° of latitude north or south of the sun. New York is about 40½° north of the equator and 64° north of the Tropic of Capricorn (40½° plus 23½°).

In midwinter it thus receives 44 per cent as much radiation as it would if the sun were overhead.

It is the way in which the amount of radiation reaching a place varies with the height of the sun which causes the temperature to vary between summer and winter, and *not* the varying distance of the earth from the sun which, in fact, has very little effect. The earth is actually closest to the sun on about 4th January in each year, in the middle of the northern winter, and this point on its orbit is called *perihelion*. The opposite point, when the earth is furthest from the sun, is called *aphelion* and is reached during the northern summer, on about 8th July.

Reverting to our experiment, if we turn the card through 90°, it is obvious that no light will fall on its surface. To the fly on the card the light will be level with its horizon. In the same way the sun cannot shine on any part of the earth which is 90° (or more) of latitude away from it. When, for example, the sun is over the Tropic of Capricorn, it cannot 'rise' in any place north of 66½°N. This gives us the *Arctic Circle*, 23½° from the North Pole. (Figure 4.) When the sun is over the Tropic of Cancer similar arguments fix the Antarctic Circle, 23½° from the South Pole. Reference to Figure 4 will show that, as the earth rotates, no point within the Arctic Circle can move into sunlight at the time of the winter solstice, neither can any point within the Antarctic Circle move into darkness. Six months later the rules are reversed.

At the actual poles of the earth the sun is above the horizon for exactly six months of the year and below it for the other six, thus giving six months' day and six months' night. During the summer the sun goes right round the horizon in each twenty-four hours, as the earth rotates, and so gives rise to the phenomenon called 'the midnight sun', which may be seen from anywhere inside the Arctic or Antarctic Circles at some time during their summers.

Let us now imagine the earth moving in its orbit round the sun, travelling at an average speed of about 18½ miles a second, or roughly 65,000 m.p.h. It takes approximately 365¼ days for the journey of a little under 600 million miles. The more precise figure is 365.2422 days. This is allowed for by having a leap year every fourth year, but omitting it every hundredth year, unless the number of the century happens to be divisible by four. (Thus 2000 A.D. will be a leap year; 2100 A.D. will not.) The cumulative error in this system amounts to less than a day in 3,000 years.

If we were to place a lamp in the middle of a room to represent the sun and walk around it, the objects which appear immediately beyond the lamp would continually change. At one moment the lamp might appear to be in line with the fireplace, the next moment with the door, then with a picture, and so on. Exactly the same effect occurs as the earth moves around the sun in its orbit. If we were able to see the stars beyond the sun, the sun himself would appear to be continuously changing his background just as did the lamp.

We do not, of course, feel the earth moving and so the sun would *actually appear to move* against the starry background. Year after year he would trace out the same track amongst the constellations, a track which was certainly known to the Babylonians 5,000 years ago, and which we call the *ecliptic*. It passes through the twelve constellations of the zodiac. When the sun has any particular constellation as a background we say that he is 'in' that constellation.

In just the same way as we look outwards away from the sun into a night sky, the stars which we see will depend on where the earth is in its orbit or, in other words, on what season it happens to be. Looking once more at Figure 4 we see that in the northern winter the Twins are visible in the night sky whereas in summer they would lie beyond the sun and so would be up only in the day-time. As the earth moves round the sun through the seasons there is a continual change going on in the constellations that can be seen. In Chapter IV we shall be looking at the stars in more detail and shall be discussing the patterns to be looked for at different times of the year. It is this ever-changing panorama of the night sky which can make our watch so fascinating.

PRECESSION

So far we have only looked at the two movements of the earth which primarily affect our daily lives; its rotation on its axis and its revolution around the sun. The other movements are less obvious but nevertheless concern the astronomer and have some interesting effects.

Because its axis is inclined to the plane of its orbit round the sun, the earth behaves rather like a spinning top that is not quite upright. We have all seen such a top and know how it wobbles slowly round and round, the direction in which its axis points appearing to trace out a circle in the sky. This movement is called *precession*. The spinning motion of the top causes it to try and point continuously in the same direction in space, as will any freely spinning gyroscope, while at the same time the gravitational pull of the earth tries to upset it, and the conflict between these two forces causes the axis to 'wander'.

The spinning earth is like the top whilst the gravitational pull of the sun on its equatorial bulge acts as the unbalancing force. The result is that, just like the top, the earth's axis also traces out a circle in the sky, taking approximately 25,800 years to complete one revolution. (Figure 5.) The pull of the moon on the earth's bulge has a similar but very much smaller effect, which varies over a period of about nineteen years and causes the earth's axis to 'nod' slightly to and fro as it precesses, so that the pattern traced out in the sky becomes a slightly corrugated circle. This effect is known as *Nutation*.

We know that the axis of the earth points very nearly at the star Polaris, which we call the 'Pole Star' and regard as the one 'fixed' star in the heavens. However, a glance at Figure 5 shows that Polaris has not always been in this favoured position. Five thousand years ago, when the Great Pyramid was being built, it was the star Thuban (Alpha Draconis) in the constellation of the Dragon, to which men referred for their direction. Passages have been found in some of the pyramids which, according to writings of the time, were orientated on Thuban. Three thousand years hence it will be the star Er Rai (Gamma Cepheii) in the constellation of Cepheus. At times the wanderings of the pole are

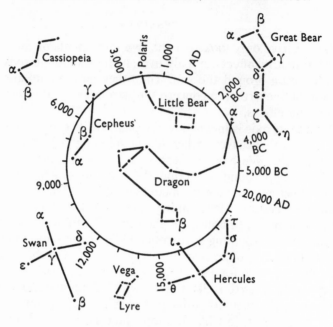

Fig. 5. The effect of precession on the ' North Pole Star '.
The complete cycle takes about 25,800 years to complete.
It will be seen that for long periods there is no bright star
to mark the position of the north celestial pole.

such that for long periods no really bright star is sufficiently near
the pivot of the heavens to act as a convenient ' Pole Star '. Shakespeare's Julius Caesar said:

> *But I am constant as the Northern Star,*
> *Of whose true fixed and resting quality,*
> *There is no fellow in the firmament*

but we cannot be certain to which star Shakespeare was referring. The ' Northern Star ' of his day was probably Polaris,
although not as well placed as it is to-day, but Polaris is unlikely
to have been so regarded by Julius Caesar 1,650 years earlier.
Indeed at that time there was no prominent star near the pivot of
the northern sky, just as to-day we find that the region surrounding the south celestial pole is quite barren of bright stars.

We are indeed fortunate in the northern hemisphere, for Polaris is now (1966) only 1° from the true north pole, and this distance is still decreasing. By 2095 it will be less than ½°.

Referring once more to Figure 4, we can discover what effect precession will have on the stars that are visible at any particular season. We see how in the position marked 'Summer Solstice' the earth's axis is inclined to the left. The constellations of Sagittarius and Scorpio are well seen in the night sky and those of Taurus and Gemini will not be visible because they will be in the same direction as the sun. Some 13,000 years hence, however, the earth's axis will have precessed through half a revolution and will be tilted in the opposite direction. Summer and winter will have been reversed, and Sagittarius and Scorpio which now grace our summer skies will be regarded as stars of the winter-time. By A.D. 15,000 the summer stars will be those splendid constellations of Taurus, Gemini and Orion which now shine down on us at Christmas-time.

Owing to precession, the constellations are gradually slipping round all the while. The change is very slow, amounting to only 1° in 72 years, and this makes little difference to the appearance of the sky during the span of a life-time. Nevertheless it was sufficiently obvious to have been noticed by the great Hipparchus in about 125 B.C. when he compared his own observations with those of early astronomers. Hipparchus called this phenomenon the *Precession of the Equinoxes* although he could not have been aware of the true reason for it. The position in which the sun crosses the celestial equator at the spring equinox is called the 'First point of Aries', and in Hipparchus' time it really was in the constellation of Aries, the Ram. To-day, however, owing to precession, it has slipped into the next-door constellation of the Fishes, and after 13,000 years it will have slipped through a complete half circle and be in the constellation of the Virgin.[1]

The wanderings of the pole introduce another interesting effect by altering the stars visible from different latitudes. Thus six thousand years ago Polaris would have been visible from much of Australia, while the Southern Cross and the Centaur

[1] This fact is seldom appreciated by those who write horoscopes!

were visible from Greece and from almost all the North American continent. This explains why many southern constellations, which are no longer visible from the Mediterranean, nevertheless have classical names.

THE CELESTIAL SPHERE

Although we know that the heavenly bodies are at vastly differing distances it would obviously be difficult to represent them in this way on star maps or star globes. On the other hand it is a comparatively simple matter to imagine them as the ancients did, all at the same distance and dotted about on the inside of a hollow sphere with the earth at its centre. This we call the *celestial sphere*. The earth's axis, extended, meets it at the *celestial poles*, and half way between them is the *celestial equator*. To define the precise position of a place on the earth's surface, we quote its latitude and longitude. On the celestial sphere we use very much the same system except that latitude and longitude are replaced by the terms *declination* and *right ascension*.

The declination of a star is measured north or south of the celestial equator in exactly the same way as is latitude on earth. However, not having a Greenwich in the sky, we have to choose another arbitrary point from which to measure right ascension (or longitude) and the point selected is the vernal equinox. Right ascension is always measured *eastwards* from the vernal equinox and is generally expressed in hours and minutes, from 0 to 24 hours. The right ascension of the star Vega is thus 5 hours, 24 minutes, and its declination is 38° 45′N.[1]

Since the wandering of the pole affects both the position of the celestial equator and the vernal equinox, it will slowly alter the declination and right ascension of every star in the sky. The date for which a star map is drawn (called its ' epoch ') is therefore always stated, and the position of Vega given above should be correctly written as 5 hrs. 24 m., +38° 45′ (1950).

Thus we have seen that the movements of the earth affect the

[1] It is more usual to write ' + ' for north, and ' − ' for South.

apparent positions of the stars in many different ways. Rotation gives us rising and setting; revolution round the sun changes the constellations visible during each season; precession and nuta- tion alter the position of stars on the celestial sphere and thus change not only the stars associated with the different seasons, but also those which can be seen from different parts of the earth.

Those who study the subject more deeply will find that as the years roll by many other changes are taking place as well. The gravitational pull of our brother and sister planets causes fluc- tuations in the earth's movement, known as *perturbations*. The shape of the earth's orbit and its orientation in space are altering, as also are the earth's speed of rotation and its rate of revolution round the sun.

These minute changes are beyond the scope of this book. Nevertheless we have already seen sufficient to realise how in- volved are the movements of the earth on which we live, even when considered simply as a planet. To these must also be added its movements through space as it attends on the sun; but these will be considered in later chapters.

HOW OLD IS THE EARTH?

In our final chapter we shall be discussing one of the funda- mental cosmological problems, the age of the universe, but a brief word at this stage about the age of the earth itself might not be out of place.

Not many years after Galileo had made his far-reaching con- tributions to our knowledge, a famous Irish archbishop named James Ussher made the rather surprising announcement that he had fixed the date of the 'creation' as 4004 B.C. He had arrived at this figure by simply adding up the ages of all the patriarchs! Would that it were all so easy. Nowadays the generally accepted age for the earth is about four and one-half thousand million years, depending on the method used for its determination.

The most recent estimates have been based on a study of the radio-active elements in the earth's crust, chiefly uranium and thorium, which disintegrate very slowly and have as their end products helium and a special form of lead, which can quite easily be distinguished from 'ordinary' lead. The rate of decay appears to be absolutely constant for each element and is un-affected by temperature, pressure, or any other physical con-ditions to which it is subjected, and which are bound to have changed considerably since the earth was formed. They thus represent an extremely accurate clock for fixing the time when the earth's crust was laid down. By measuring the proportion of lead 'isotopes' and of helium in a suitable mineral deposit, it is possible to determine how long the process of disintegra-tion has been going on or, in other words, when the clock was started, and this has led geologists to the conclusion that the age of the earth is about four and one-half thousand million years.

Other estimates have been made, based on different premises, which add weight to this figure. For example, by measuring the amount of salt carried down to the sea each year it is possible to calculate how long it would take for the oceans to reach their present salinity. This gives the somewhat lower figure of 1,500 million years, which is of the order we should expect, as the seas must certainly be much younger than the earth itself.

The conditions which obtained on earth 'in the beginning' will have depended on how it came into existence, but were clearly very different from those we know to-day. Nevertheless, there are indications that life in some form has existed for quite a large proportion of the total life of the earth, and a likely time-scale might run something like this:—

> Age of the earth—about 3,000 million years
> Existence of life—nearly 1,000 million years[1]
> Existence of 'man'—about 1 million years[1]

Why life started when it did we do not know, but the chances that conditions might develop which would be suitable for it are perhaps rather easier to assess. They will be discussed in greater detail in Chapter IV, when we have considered what types of star might make suitable parents for families of planets like our own.

[1] Recent experiments have shown that both these figures may have to be considerably increased.

THE MOON—OUR NEAREST NEIGHBOUR

Of all the heavenly bodies the moon is perhaps the one whose study is most rewarding to the amateur observer. We know her as the fair and friendly lady whose gentle light softens the darkness. The dark patches on her surface, visible even to the naked eye, we know as 'The Man in the Moon'. A glance with a pair of binoculars is sufficient, however, to show that the moon is neither fair nor friendly, for the rugged mountain ranges and the thousands of craters that pit her scarred surface can be seen standing out starkly against the smooth dark plains. We realise at once what a barren and inhospitable world this neighbour of ours is.

The moon has a diameter of 2,160 miles, a little more than a quarter of that of the earth. The volume of fifty moons could thus be fitted inside our planet. However, the material of which she is made is less densely packed than in the earth, and it would take eighty-one moons to 'balance' one earth. Her 'density' is thus only about $\frac{3}{5}$ of that of the earth. Because of her elliptical orbit, the distance of the moon varies considerably between about 226,000 and 252,000 miles, with a mean value of 238,857, which can be determined very accurately by using the method usually employed by surveyors. When a surveyor wishes to find the distance of some inaccessible peak or off-shore island he carefully measures out a baseline and observes

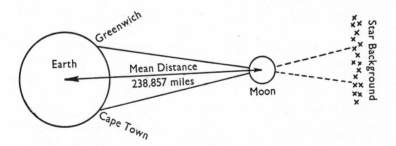

Fig. 6. Measuring the distance of the moon by the trigonometrical method. The moon is photographed simultaneously against the stellar background, from two observatories a long way apart.

the object from either end. He is thus able to plot its position and measure its distance off his map. Exactly the same technique can be employed with the moon, although in this case a base line of some thousands of miles is needed. Two observatories in different parts of the world observe the position of the moon photographically against the background of distant stars at precisely the same instant and, from the apparent difference in her position, her distance can be obtained to within one mile. (See Figure 6.) More recently this distance has been confirmed by direct radar measurements. In round figures, however, we will call it 'a little less than a quarter of a million miles'. Although the 1,000 m.p.h. jet aircraft we talked about in Chapter I would take about ten days to make the journey to the moon, the American Mariner space-craft which in 1965 gave us the first close-up pictures of the moon's surface took only thirty hours. Already the moon seems to be almost within our reach.

By far the most obvious feature of the moon's behaviour is the way she goes through her regular series of phases each month, yet these are often very incompletely understood. If we hold a tennis ball in the beam of a flashlight, only half of its surface will be illuminated. Depending on where we stand with respect to the tennis ball we shall be able to see either the whole of the illuminated half, or a portion of it or, when we are on the opposite side of the ball to the torch, none at all.

This is precisely what occurs with the moon, as is illustrated in Figure 7. When the moon is between the earth and the sun the illuminated half is turned away from us, and this is the moment of 'astronomical new moon'. Gradually as she moves away from the direction of the sun, we begin to see a thin crescent of light hanging low in the west after sunset. This is often referred to as 'the new moon' — and we turn our money or wish, depending on our choice of superstition — but by the time she is visible to the naked eye the moon is probably at least a day and a half 'old'. Each night we see a little more of her surface light up as the *terminator*, the shadow line between the sunlit and darkened parts, creeps across her from west to east, showing the sun rising on more of her surface; each night she will appear a little further to the east amongst the stars as

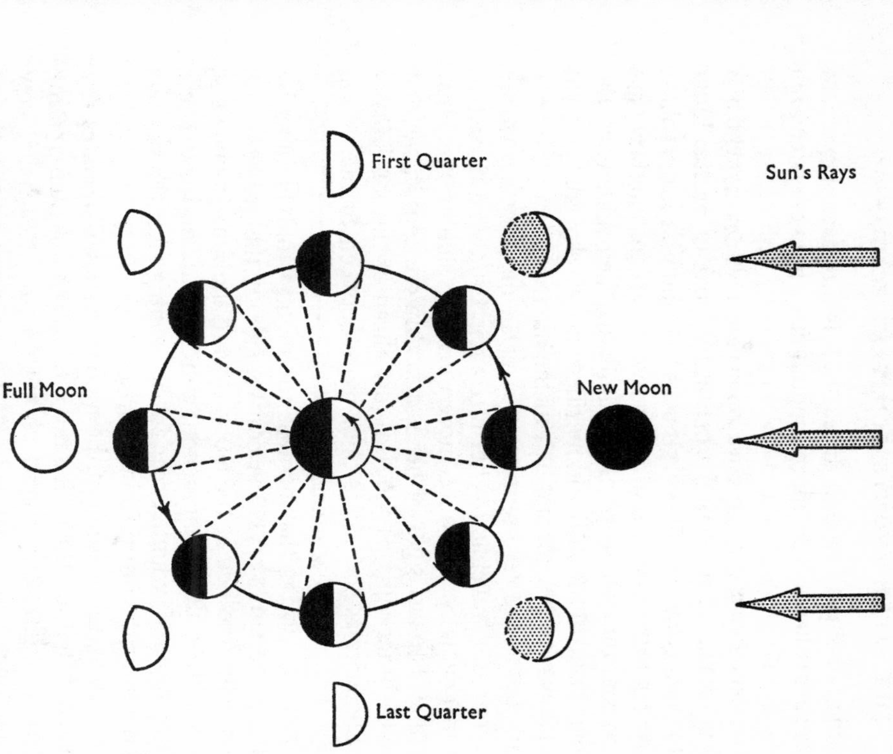

Fig. 7. The Phases of the Moon. As the moon travels round the earth more and more of the illuminated hemisphere can be seen until, about 14½ days after 'new moon', the full illuminated disc can be seen. Thereafter the visible part of the illuminated hemisphere decreases until the moon again passes between the earth and the sun at the next 'new moon'. When the moon appears as a crescent the dark part is illuminated by earthshine, giving the appearance known as 'the old moon in the new moon's arms'.

N.B. As seen from the earth (centre of figure), the 'first crescent' and 'first quarter' are bowed to the right (as a 'D') whilst the 'last quarter' and 'last crescent' are bowed to the left (as a 'C').

she pursues her path around the earth. It is this latter movement which causes her to rise approximately fifty minutes later each night.

After roughly seven days we observe a half-moon, which is known as the 'first quarter' because a quarter of the lunar month has gone by. The 'gibbous' or hump-backed phase follows, and the moon continues to wax until at length we are able to see the whole of the illuminated hemisphere as the glorious disc of the full moon rides in our sky like a great celestial mirror. The second half of the month follows the same pattern, but in the reverse sequence. As the moon wanes we see the terminator travel across her face again from west to east, as the sun gradually sets once more over our side of the lunar landscape. The last sight of her is as a thin crescent in the east in the early morning as she rises just ahead of the sun. Finally she disappears as she passes once more between the earth and the sun, and so a new *lunation* begins.

It is worth noting that because the moon is only shining by reflecting the sun's rays, the convex side of the crescent will always lie on the side of the moon toward the sun. As one child aptly described the phenomenon, a crescent moon looks like a ' smiling mouth ', whether seen in the evening after the sun sets or in the morning just prior to sunrise.

The waxing and waning of the moon are shown in Plates 1, 2 and 3. To watch her through a full lunation can be a fascinating pastime as night after night, or even hour after hour, new features come into the sunlight. Just as on earth objects throw long shadows at dawn and dusk when the sun is low in the sky, so on the moon we can observe the lengthy shadows cast by the mountain ranges and crater walls. When it first becomes visible, the inside of a crater may be completely filled with shadow but, as the sun climbs higher, so the darkness recedes across the crater floor until we can see right into its open bowl. Only as it crosses the smooth dark plains does the terminator appear as a clear-cut line. In other places, owing to the roughness of the terrain, it appears irregular and broken as the high crater walls catch the rays of the rising sun while the depressions remain longer in shadow. On occasions high peaks may suddenly light up and flash out as brilliant points of light from well within the darkened sector. If we watch the landscape gradually unfolding in this way the principal features can be identified one at a time and the business of learning the lunar ' geography ', an essential first step for any observer, is very much simplified.

The best time for lunar observation is near the first or last quarter when the illuminated features are thrown into prominence against the shadows of early morning or late evening, for as the sun climbs higher the shadows recede and near the time of full moon, when the sun is ' overhead', they disappear altogether. Beautiful as the moon may look on these occasions, this is not the best time to observe her.

An exception, however, occurs in the case of the bright ray systems which stretch like tentacles from certain prominent craters, in particular Tycho and Copernicus. (See Plate 2.) They are normally invisible, but under conditions of high illumination they suddenly become very conspicuous and spread over much of the moon's surface.

Before looking more closely at the nature of the moon let us examine the way in which she behaves, for although we say that she goes round the earth once in a month this is not strictly true. According to Newton's law of gravitation every body attracts every other body, so that while the earth is holding the moon in its grip, the moon is also influencing the path of the earth. In fact the two bodies are really *revolving round each other* as they both move around the sun. The earth and the moon are locked in each other's attraction, rather like two dancers, spinning round a ballroom to the tune of an old-fashioned waltz. If the dancers are of comparable size they will each move approximately the same distance to either side of their line of advance. On the other hand if one is a stout old gentleman and his partner a fair young lady, the latter will find herself swung round and round whilst her stately partner deviates only slightly from their mean track. So it is with the earth and the moon. On this occasion, however, the earth is so very much more massive than the moon that the point about which they revolve, the centre of gravity of the earth-moon system, is about 1,000 miles *inside* the earth. This point is called the *barycentre*, and follows the true elliptical orbit round the sun, while the centre of the earth wobbles slightly from side to side, like the stout old man.

For simplicity, however, we will continue to talk about the moon revolving round the earth, a journey which takes approximately 27⅓ days.

At the time of new moon the moon is between the earth and the sun, as we have already seen. After 27⅓ days, although she will have arrived back at the same position in the sky relative to the background of stars, the earth will also have moved on in its orbit, so the moon will have to travel a little further in order once again to be between the earth and the sun. Thus although the moon goes round the earth in 27⅓ days, the interval between one new moon and the next becomes approximately 29½ days, which we call a lunar month.

It is well known that the moon always keeps the same face turned towards the earth. What this means, in effect, is that as she travels once round the earth she must also turn once on her own axis. At first sight it would appear to be a remarkable coincidence that the two periods should so exactly coincide, but the reason is comparatively simple. The moon, as we are aware, raises tides in our oceans by its gravitational attraction. To a lesser extent so, also, does the sun, but if for the moment we ignore the sun's influence and consider only the moon, it will be apparent that the peak of the tide she raises will always be directly underneath her. As the earth rotates, this peak moves over the surface, giving rise to alternate high and low tides.[1] The flow of water round the earth caused by the tides exerts considerable friction, and this acts as a braking force on the earth's rotation. Consequently the earth's rotational energy is gradually being used up, and its rate of spin is being slowed down. The length of the day is increasing by about a thousandth of a second every century. One day the moon will win, and she and the earth will then go round rather like the dancers, always keeping the same face towards each other — but this is unlikely

---

[1] Gravitational forces pull the water away from the earth on the side toward the moon. Similarly they pull the moon away from the water on the opposite side of the earth, so that the water boundary of the earth is slightly egg-shaped, giving two high tides and two low tides in each twenty-four hours.

to happen for about fifty thousand million years.

In just the same way the earth raises tides on the moon. Even though the moon is not covered with water the same braking force of tidal friction is at work on the rocks of her surface. Gradually, through the ages, the rotation which she must once have had has been slowed until the 'bulge' pointing towards the earth has finally stopped, held tightly in the earth's gravitational grip. She can thus only turn on her axis in the same time that it takes her to circuit the earth, and we on earth can only study the features on the side that is turned towards us.

There is, however, one slight reprieve. The moon turns on her axis at a constant speed, but, because of her elliptical orbit, her speed around the earth varies. We are thus enabled to see slightly 'round the corner' from time to time. Also, because the orbit of the moon does not lie precisely in the same plane as that of the earth, but is inclined to it at an angle of approximately 5°, we can sometimes see a little further across her north or south poles. These effects are known as *libration* and result in some 59 per cent or roughly four-sevenths of the moon's surface being visible at one time or another.

The remaining 41 per cent can never be seen directly from the earth, and until the historic flight of Lunik III, launched on 4th October 1959, we had no way of telling what lay on the further side. Even now, the photograph shown in Plate 4 has many imperfections as it was taken under conditions of high illumination so that the features do not stand out well. Nevertheless a comparison with Plate 2 shows at once that some of the dark plains belonging to our side of the moon are clearly visible on the left-hand edge of the photograph, so that we can be fairly certain that this type of feature is less abundant on the far side. In assessing the merits of this picture we should perhaps remind ourselves that this was the first of such photographs ever to be taken. Naturally it could not be taken through a powerful telescope, as would be used from the earth, and when we remember that the result had to be televised back across about 300,000 miles of space, equivalent to going twelve times round the world, it must be regarded as a remarkable achievement by any standard.

EXPLORING THE MOON

Supposing that we are already able to land on the moon, suitably protected and equipped for carrying out exploration, in what sort of a world should we find ourselves? The scene before us would be one of utter desolation, probably largely in a shade of grey-black. We could hear no sound, for the moon appears to have no appreciable atmosphere, and sound waves require some sort of medium in which to travel. Our new abode would thus be not only barren but also completely silent.

Let us assume that we have landed on one of the great plains, known as the 'Seas' or the 'Maria', because at one time the early astronomers believed that the moon possessed oceans and continents just as does the earth. Although we now know that there is apparently no water on the moon, it is possible that at some time in the past the 'seas' did contain some form of liquid lava which solidified into the vast, smooth plains we now see In places the tops of crater formations or mountain peaks can still be observed breaking through the surface, while the walls of many craters have clearly been breached by its onslaught. Because of the moon's curvature our horizon on the plains would be a mere mile and a half away.

Some astronomers have suggested that the lava surface of the 'seas' is covered by a very thick layer of dust or volcanic ash, produced not by eruption but by the continual bombardment of the lunar surface from outer space. The remarkable photographs taken from the U.S. Ranger lunar probes, a few miles before they crashed into the lunar surface, confirms the 'soft' character of the surface. However, there is considerable doubt that the heavily fragmented layer goes down more than a few feet.

Measurements taken from the earth have shown that the moon's surface undergoes tremendous changes of temperature. With the sun overhead the rocks would be quite unbearable to the touch, with a temperature of approximately 100° C., about the temperature of boiling water at the surface of the earth. As darkness falls the heat radiates away into space very rapidly, for there is no blanket of air to retain it, and the temperature drops to below zero in less than an hour. During the long lunar night, which lasts approximately two weeks, it may well fall to about $-130°C.$, or the equivalent of what we on earth would call 200° below zero on the Fahrenheit scale. If, however, we were to burrow into the dust or ash of the plains we should probably find that only a few inches below the surface the temperature reached a fairly stable value of about 0°C., near the freezing point of water, because the layer of dust or volcanic ash is such a poor conductor of heat. As a consequence, this temperature would be very little affected by the tremendous changes occurring on the surface only a few inches higher up. It is clear that neither insulation nor refrigeration will present any problem on the moon!

What else will be found below the surface is still a matter for speculation. It has been suggested that there may well be rich mineral deposits including uranium, that moisture may be locked in the rocks as water of crystallisation, or, protected by an insulating layer of dust, may exist as a layer of ice, waiting to be mined. Lunar explorers would obviously be unwise to count on finding water quickly, but such discoveries would clearly prove very valuable. At present, however, it appears that the deserts are remarkably uninviting, and it is interesting that they should have been given such delightful names—' Sea of Rains ', ' Sea of Serenity ', ' Sea of Tranquility ', ' Sea of Fertility ', ' Bay of Rainbows ', ' Lake of Dreams ', ' Sea of Nectar ', and so on. All give the impression of peace and well-being. Only the ' Sea of Crises ' and the ' Ocean of Storms ' seem to be out of company! Most of the seas appear to be connected to one another in just the same way as are the oceans of the earth, suggesting that they may all have been formed at one time from the same great lava flow.

If, instead of landing on an open plain, we had chosen to touch down inside one of the smaller craters, we should find ourselves hemmed in on every side by rough, forbidding slopes, thrusting their jagged peaks into the jet-black sky above. We have selected a *small* crater, because if we had chosen one of the larger ones its floor might well have been indistinguishable from the surface of a plain! From the middle of such great craters as Plato or Archimedes the walls would be quite invisible as they would lie below the horizon on all sides, even though they are thousands of feet high.

Roughly 10,000 craters have been mapped on our side of the moon, varying in size from pin-pricks, a mere few hundred yards across, to the great walled plains more than 100 miles in diameter. Some appear quite smooth inside whilst others have massive central peaks rising from their floors. About five hundred have been given individual names, mostly after great men of science. The ancient Greek astronomers Archimedes, Pythagoras and Ptolemy thus have their memorials alongside the giants of the renaissance, Copernicus, Tycho Brahe, Galileo, and Newton. It is encouraging to note that the Russians have perpetuated this idea when naming the features on the further side of the moon (see Plate 4), but it is to be hoped that they will retain an international flavour, for there is a limit of the number of tongue twisters the astronomer can be expected to master!

The largest crater visible is Clavius, near the south-eastern rim of the moon. This is a great walled plain about 158 miles across, and its area could comfortably accommodate the whole of Wales. Smaller but more conspicuous are the craters Copernicus and Eratosthenes in the Sea of Rains, the great trio Ptolemaeus, Alphonsus and Arzachal in the centre of the disc, and Tycho which, under conditions of high illumination, displays his brilliant ray system and becomes the brightest object on the moon.

The three and a half centuries of telescopic astronomy have produced numerous ingenious theories to explain how this jungle of craters came into existence. Many, however, have displayed rather more imagination than sound astronomical knowledge, so that at present there are basically only two rival hypotheses which merit our attention. Both have their unswerving adherents amongst our eminent astronomers! Although it is certain that the craters of the moon cannot be compared with volcanoes on earth, the idea that they were formed by some type of volcanic action is obviously very attractive. One idea suggests that there may have been some form of 'bubbling' when the moon was in a hot, plastic state and that the craters represent the rims of the burst bubbles; another suggestion pictures narrow fountains of lava spraying outwards to form the rings. Either explanation might be possible and there are many other variations on the same theme. The fact that the walls of many of the large craters have smaller and obviously younger craters breaking through them suggests a gradual slackening of activity, and lends weight to the 'volcanic' theory.

The rival theory is very different. It suggests that the craters were caused by an intense bombardment of the moon's surface by giant meteorites in some bygone age. If we watch a small boy throw a handful of pebbles into the sand from which the sea has just retreated, the pattern produced is very reminiscent of the craters of the moon. Experiments which involve firing objects at very high speed into plaster of Paris have produced patterns which are even more convincing. The exponents of the meteoric theory point to the marked similarity between the lunar craters and certain cavities on earth, notably in Arizona (see Plate 5) and near Lake Ungava in North Canada, which were certainly formed by the impact of meteors. Most astronomers would agree that the great majority of the lunar craters, especially the large ones, were caused by meteoric impact. However, there are many secondary craters produced by matter ejected from the parent crater at the time it was formed. These are the so-called 'splash craters', which often are elliptical or almost linear in shape. Also, the Ranger photographs have clearly shown that long lines of craters exist, which clearly have arisen from some form of associated volcanism.

Apart from the 'seas' and the craters, there are numerous other features on the moon which can be picked out with the help of a small telescope. There are the towering mountain ranges which border many of the seas, in particular the Alps the Apennines and the Carpathians surrounding the Sea of Rains (Plate 3) and in the south, the ragged Leibnitz mountains whose tallest peaks would tower 6,000 feet above Mount Everest. Among other configurations that can be identified are the clefts and valleys which cut into the mountains, the Straight Wall, 60 miles long and 800 feet high, the sharp hummocks and isolated peaks which rise from the plains and, almost everywhere, the delicate tracery of the rilles.[1] Finally there are the mysterious rays, best seen near the time of full moon, which spread in all

[1] Fine cracks in the surface.

directions from Tycho in the south, from Copernicus and from about a dozen other less important craters.(Plate 2.)These strange streaks probably consist of layers of fine dust ejected from a crater at the time it was formed by meteoric impact. Indications are that the rayed craters may be somewhat younger than the others. For example, the ray system from Tycho clearly covers many old and partially ruined craters.

The lunar explorer will find no lack of variety, just as to-day the explorer with a telescope can experience the fascination of studying the countless features under every variation of light and shade.

Although we think of the moon as a dead world there have been indications that she may not be as 'dead and unchanging' as was once believed. In November 1958 and again in 1959, the Russian astronomer Kozyrev reported seeing an unexplained reddish glow in the crater Alphonsus. Although originally reported as a volcanic eruption, it was probably an eruption of gas containing some form of carbon, since the bands of the carbon molecule clearly appeared in the spectrum. This is why Alphonsus was the target of the successful shot Ranger IX.

When we start to explore the moon's surface one great advantage will immediately become apparent. Because she is such a small world the gravity at her surface is only about one-sixth of that at the surface of the earth. We would thus seem to have only about one-sixth of our earthly weight on the moon, and movement across her surface would be greatly simplified. A man who could jump six feet high on the earth could break all records with a leap of over twenty feet[1] on the moon; an Olympic long jumper could clear half the length of a football field. We need therefore have no fear about leaping over precipices twenty or thirty feet high for we should float comparatively gently to the ground, and mountain climbing will be very much simplified.

To date, all efforts to detect any atmosphere on the moon have

[1] Note that he could not jump *six times* as high on the moon. When a jumper clears 6 feet he probably only raises his centre of gravity about three feet from its original height of 3½ feet above the ground. On the moon, therefore, he could jump about $(6 \times 3) + 3\frac{1}{2} = 21\frac{1}{2}$ ft.

failed, and its absence is another direct result of this low gravity. We all know that when we on earth throw a ball into the air the force of gravity quickly pulls it back again, but that the harder we throw it the higher it gets before this occurs. If we could give it a speed of seven miles per second it would never come back and, in consequence, this is called the *velocity of escape*. It is the speed that must be given to a rocket or to any other object to enable it to break clear of the earth's influence. The atoms and molecules in the air do not often reach this speed, so that the earth retains its atmosphere intact. On the moon, however, owing to the low gravity the velocity of escape is only about one and a half miles per second, a speed that can certainly be reached by the atoms of most gases, particularly under the conditions of extreme heating that occur on the moon's surface. So although it seems likely that the moon once had an atmosphere, it cannot have been long before it leaked away into space. On the moon we shall thus enjoy none of the advantages that our atmosphere bestows on earth. Standing on the moon's surface we shall be virtually standing ' in space '. Apart from having no air to breathe, we shall have lost our equitable climate, our blue sky and the protection from meteorites and the many harmful cosmic and ultra-violet rays. Even long-range radio communication will be impossible without a radio mirror in the form of the ionosphere, and it will be necessary to employ communications satellites or perhaps even to bounce our messages off the earth!

On the other hand without the interference of an atmospheric blanket the radiation from the stars and the distant galaxies will reach the moon's surface in the same untarnished state in which it started its journey, perhaps millions of years before. The advantages to astronomy of setting up observatories on the moon must surely be obvious to all.

The absence of atmosphere coupled with a complete lack of water indicates that, apart from great heat and cold, there can be very little in the way of ' weather ' on the moon; certainly no wind or rain. On earth the wind and the rain, the snows, the glaciers, even the movements of the oceans, are all continually changing the surface features. Year after year, century after century, the peaks are worn down, the cliffs are cut away and gradually all trace is lost of what went before. None of this weathering can take place on the moon. Although the enormous changes in temperature on her surface undoubtedly cause much erosion. Its work must be comparatively slow. The major erosive agent on the moon is undoubtedly the influx of meteors, large and small. The larger ones may have produced major Maria such as Mare Imbrium. Splash craters appear to have spread out in all directions from this area. In one place mountain ranges have been sharply cut to form the Alpine Valley. Smaller meteors may even chip off lunar fragments that ricochet back into space, going into orbit to form additional meteors.

The remarkable series of photographs taken from the highly successful NASA lunar probes, Ranger VII, VIII and IX, have revealed much information hitherto unknown about the fine structure of the lunar surface. Earthbound photographs, even by the largest telescopes, are always blurred by scintillation of the earth's atmosphere. The Ranger photos, taken in the vacuum of interplanetary space, display no such distortion. The last photograph taken, just before Ranger IX crashed into the surface, showed small craters no bigger than ten inches in diameter. Indeed, the entire lunar surface seems to be heavily cratered, the result of both primary impact and secondary splash craters, in addition to the craters of volcanic origin. The lunar rilles have also proved to consist of lines of craters, evidently associated with long cracks in the lunar surface.

No wonder men are excited at the prospect of going to the moon. If scientists could study her surface at first hand the answers to so many of the puzzles that to-day confront them from a quarter of a million miles away would surely become apparent. Clues might be found which would help to solve the mystery of her origin, possibly also of the origin of the earth, and with that may follow the answer to how the solar system itself came into existence. The moon holds the key and the present generation will surely see the door unlocked.

CHAPTER III

—••E)(3••—

THE SUN AND HIS FAMILY

IN OUR opening chapter we built up a model of the solar system in which the sun figured as a three-foot globe controlling by its gravitational pull an assortment of planets spread over an area roughly five miles across. As we look more closely at this little community in space we shall find that as individual worlds the planets differ from one another even more widely than they did in our model. However, before we pay each one a visit, let us start by examining the sun himself, who possesses 99·9 per cent of all the matter in our solar system and so completely controls the movements of his far-flung family.

The sun is a great ball of gas with a diameter of approximately 864,000 miles. Yet the clear, brilliant, round disc which we see, called the *photosphere*, is really something of an illusion. Already we have noted that the earth is immersed in the sun's atmosphere, but since the sun is gaseous all the way through, the boundary cannot be as sharply defined as it can with a solid body. If we were to drop an imaginary indestructible cannon ball into the sun it would penetrate millions of miles into the outer corona and probably many thousands of miles into the photosphere before encountering any appreciable resistance. When we look at the sharply defined disc of the sun it is hard to realise the extremely rarefied nature of its outer regions. Apart from the question of the intense heat, an astronaut flying into the sun would be quite unaware of the fact that he had penetrated into the interior.

Our sun is a remarkably constant furnace, and his output of light and heat has not fluctuated up or down by more than 2 per cent since measurements have been possible. This is very fortunate for us, for if he were to vary in the same way that a great many of the stars are seen to do, life on earth would become very difficult if not impossible. For example, it would require only a comparatively minor increase in the sun's output to melt all the ice and snow around the poles of the earth. The general level of the oceans might then rise by approximately a hundred feet and the habitations of over a quarter of the world's population would be submerged. In the other direction, if the sun's output were to fall off appreciably we should very quickly freeze. It is still a matter for conjecture whether the great ice ages of the past may not have been due to unexplained variations in the amount of the sun's radiation reaching the earth.

Astronomically the sun is of tremendous importance to us. Being easily our nearest star (the next is over a quarter of a million times as far away), he is the only one which shows us a disc and whose surface detail we can study. In a sense we can examine the sun under a microscope and yet, splendid as he may appear, we must constantly bear in mind that our specimen is really only a relatively minor and rather cool example of the stellar population. His behaviour is thus likely to be comparatively mild as stars go and generally speaking we should regard him as a respectable middle-aged citizen, probably between 4,000 and 5,000 million years old.

How big is this sun in terms of the earth? It would take 109 earths placed side by side to stretch across his diameter and roughly a million and a quarter earths, packed in tightly, to fill his volume. However, being completely gaseous, he is on the average only a quarter as dense as the earth and therefore ' weighs ' only about a third of a million times as much. At his ' surface ' the gravity of the sun is roughly twenty-eight times that of the earth so that the average man (if he could survive!) would weigh about two tons. The velocity of escape is 384 miles per second (compared with 7 m.p.s. on the earth) so that the tremendous heat must cause the atoms and electrons to move at incredible speeds. This is particularly true in the upper solar atmosphere, where shock waves heat the gases to temperatures of more than a million degrees. Some of this material leaves the sun and flows out in all directions through the solar system to form what is now generally called 'the solar wind'. Many of the space probes sent out by the U.S.A. and the U.S.S.R., but especially the famous probe to Venus, recorded the existence and speed of the solar gases. The intensity of the solar wind is not constant, but varies with the stage of solar activity.

INSIDE THE SUN

The temperature of the sun's 'surface', the photosphere, is about 6,000° centigrade, which, by stellar standards, is regarded as comparatively cool. If, however, we were able to penetrate into the sun's interior, we should find that the temperature rose very rapidly indeed until, at his centre, it reached a value of about 20,000,000° centigrade. Here the atoms are completely stripped of their electrons and are probably packed very tightly together under the tremendous 'weight' of the material above them. The pressure in the middle of the sun is likely to be equivalent to several thousand million atmospheres, and the matter in the average human being would be compressed to about the size of a raspberry.

It is in this region, under these fantastic conditions of temperature and pressure, that the nuclear reactions which give rise to the sun's output of energy can take place. Here, by a process similar to the fusion that occurs in a hydrogen bomb explosion, hydrogen, the principal stellar fuel, is being gradually converted into helium and, in the process, releases 'energy'. The sun is slowly destroying himself in order to provide his fantastic output of radiation, and is 'losing weight' at the rate of about *four million tons* every single second. Fortunately for us he is so vast that he will comfortably be able to continue at this rate for some thousands of millions of years without any noticeable difference.

The enormous forces at work within the sun maintain a very neat balance amongst themselves. The gravitational forces would tend to cause the outer layers to collapse inwards, but this is largely counteracted by the tremendous outwards pressure of the gas inside which is trying to expand. In addition, as the incredible quantity of energy liberated in the sun's centre forces its way outwards it, too, exerts an outward pressure, which is known as *radiation pressure*. All these forces are mutually in balance so that the sun remains stable and neither collapses nor puffs himself out. Nevertheless the surface of the sun is violently disturbed and a closer look at his serene disc shows it to resemble a great bubbling cauldron, for ever boiling up and erupting, and always in tremendous turmoil.

WHAT IS THE SUN MADE OF?

After the telescope itself, perhaps the most powerful tool for studying the nature of the heavenly bodies is the spectroscope, an instrument which splits up light into its component colours, spreading it out to form the ' spectrum ' in exactly the same way as raindrops give rise to a rainbow. In 1814 a German scientist named Fraunhofer discovered that the spectrum of the sun was crossed by a number of dark lines which have become known as the *Fraunhofer lines* and have proved of vital importance in studying the chemical composition of stars.

In its simplest form this is done in the following way. An atom of any substance can be pictured as having a nucleus of protons and electrons, with a number of other electrons moving in ' orbits ' round it rather like a solar system in miniature. Although this model, known as the *Bohr atom*, is a gross over-simplification of the truth, it will serve our purpose. There are only certain fixed orbits in which the outer electrons can move, but under special conditions they can jump from one orbit to the next. When an atom is heated by a source of radiation it absorbs energy at a particular wave-length, and this can cause an electron to jump to a ' higher ' orbit. At the same time the radiation is robbed of some of its energy, thus leaving a gap in its spectrum *at that particular wave-length* which appears as a dark Fraunhofer line, or, as it is usually called, an *absorption line*. By experimenting with atoms of different substances in their laboratories, and noticing the positions of the absorption lines, scientists are able to identify many of the same substances in the sun and stars from the position of the dark lines in their spectra. All atoms, in fact, have a number of different lines associated with them, each of which represents a jump of one electron between a particular pair of orbits and is associated with a particular temperature. It is thus possible from a study of absorption lines not only to identify the atom but also to tell the temperature of its environment. At high temperatures the speeding electron or an energetic light ray may tear away completely an electron from its parent atom. This means that the atom becomes ionised and produces a completely new series of spectral lines. More than five thousand lines have been identified in the spectrum of the sun, and about two-thirds of the known elements on earth have been identified as existing there.

Should an atom emit energy instead of absorbing it, the electron will jump to an inner orbit, and the radiation given off appears as a *bright* line in exactly the same place. This is known as an *emission* line and can be identified in just the same way.

THE SUN'S DISC

With the help of the spectroscope it is possible not only to find out what the sun is made of, but also to photograph him in the light of a single line in his spectrum. In this way the tremendous detail of the apparently featureless disc can be brought into sharp relief. The photograph in Plate 7 was taken in the light of one of the hydrogen lines and shows that the surface is not smooth but is covered by roughly circular *granules*, each about 1,000 miles across. Each of these only lasts a few minutes and they are in continuous turmoil, probably representing the tops of convection columns that have welled up from the sun's interior.

Far more obvious features are the sunspots, which erupt in the surface and sometimes cover hundreds of millions of square miles. Occasionally they are large enough to be clearly visible to the naked eye, and would swallow the whole earth many times over. Although the Chinese annals contain many records of dark spots seen against the solar surface, they do not seem to have been recognised in Europe before Galileo turned his telescope on to the sun[1] and, by watching them as they were carried across the disc, discovered that the sun was rotating. It has since been found that, owing to the gaseous nature of the sun, different parts of his surface rotate at different speeds and, while a point on his equator travels round in about twenty-five days, a point near the poles takes some thirty-four days to complete a revolution.

---

[1] Galileo made the great mistake of *looking* at the sun with his telescope and, as a result, went blind in his old age. Amateur astronomers should always study the sun by projecting its image onto a white card and *never* by looking at it directly with any optical aid, even through filters.

Intense magnetic fields, found in the sun's atmosphere, appear to be responsible for sunspots. The magnetism changes both the pressure and temperature of the areas, so that the sunspots are cooler than the surroundings, with temperatures of about 4,000° C. They are, thus, about 2,000° C. cooler than the surrounding photosphere. The sunspots are relatively quiet areas, but the regions immediately surrounding appear to be hotter and much more active than regions without spots.

Observations over the past 150 years have shown that on the average the frequency of sunspots seems to vary with a cycle of a little over eleven years. 1958 was a year of maximum activity, when many very large spots appeared, some of which were even visible to the naked eye. The region of the sunspots appears to vary with the state of the cycle. At minimum they form mainly in intermediate latitudes, but as the cycle proceeds, new spots develop closer and closer to the equator. Very seldom, however, do they occur actually on the equator and never at the poles. After maximum the spots gradually die out and eventually a new cycle starts again in high latitudes. Thus the region of the sunspots seems to depend largely on the degree of spottiness.

The magnetic fields associated with sunspots display a regular and characteristic variation. In one eleven-year period, for example, the leading spot of a pair, in a given hemisphere, will have a North Magnetic Pole and its companion a South Pole. The reverse will hold for the opposite hemisphere. During the next cycle, however, the polarities completely reverse in both hemispheres. Therefore, the true cycle is twenty-two years rather than eleven. In addition to the intense localised fields of the sunspots, the sun possesses a general field that also slowly fluctuates, but not necessarily exactly in keeping with the sunspot cycle.

The chief effect of the sunspot cycle on the earth consists of magnetic storms, as erratic variations of the compass needle are called. Bright displays of aurorae often accompany intense, active spots. There are marked changes in the earth's ionosphere, including its reflectivity for radio waves. In addition, bright solar flares cause radio signals to fade on the daylight hemisphere of the earth. Enormous variations occur in the radiation trapped in the Van Allen belts. It has further been suggested that solar variations have an effect on the weather, but such a relationship has not yet been clearly and definitely substantiated.

The *solar flares*, previously mentioned, are brilliant, low-level eruptions in the vicinity of large and active sunspots. They are most conspicuous in the light of hydrogen, and undoubtedly represent volumes of extremely hot gas associated with the active spot. Changes of brilliance occur rapidly, within a matter of minutes. The ultraviolet radiation and X-radiation from the flare profoundly effects the lower ionosphere, as previously indicated. Such eruptions also are accompanied by bursts of short-wave radio noise.

Plate 7 also shows many bright areas called *faculae*, which surround a sunspot area. They appear well before the sunspot breaks through the surface and often remain as the surface ' heals ' after the sunspots have disappeared.

But by far the most spectacular of solar features are the magnificent *prominences*, great clouds of incandescent gas. These fall in two main classes. First of all there are those prominences that appear to form high in the *corona*, cascading back to the solar surface in graceful curves and arches. Solar magnetic fields are responsible for guiding the material. From time to time, powerful eruptions occur. Surges or twisting loops spring upwards hundreds of thousands of miles from the vicinity of active sunspots. The structure of the sun's atmosphere is enormously complex.

Studies have shown that the extreme outer atmosphere of the sun, known as the *chromosphere*, is very much hotter than the sun's normal surface, the *photosphere*. The latter has a temperature of about $6,000°$, whereas the chromosphere and associated prominences attain temperatures in the range from $10,000°$ to $50,000°C$. The solar corona, a faint pearly halo that, until recently, could be seen only at the time of total solar eclipse, is very much hotter. Coronal temperatures range between $1,000,000$ and $2,000,000°C$.

A special optical filter, invented early in the twentieth century, enabled astronomers to photograph solar flares, prominences, and other features of the sun in the light of hydrogen. But the corona could be seen only at a total eclipse until 1930, when a French astronomer named Lyot invented an instrument called a *coronagraph* which, in effect, produces artificial eclipses and makes possible systematic study of the prominences and the solar corona.

A total eclipse of the sun is, nevertheless, still a rare and wonderful sight (Plate 8) and if the opportunity to see one arises it should not be missed. As the sun's disc becomes obscured, the sky and the surrounding landscape gradually darken. The sun itself becomes no more than a thin crescent of light and, with the approach of totality, the moon's shadow can be seen approaching across the landscape with great speed. The hush of evening descends and the birds retire for the night. Occasionally, just before the moon finally obscures the sun, the solar rays shine through the lunar valleys between mountains on the edge of the moon. The resulting effect of bright patches is known as ' Baily's Beads '.

The sun then disappears and the beautiful solar *chromosphere* springs into view as a pastel pink region completely surrounding the dark disc. The faint, pearly light of the solar corona extends outwards, sometimes two or three solar diameters. The pink prominences then appear against the darker sky. Some of the brighter stars and planets shine out and even an occasional comet may appear. Then, as suddenly as it arrived, the glorious vision vanishes. The Baily's Beads appear on the opposite edge and the first thin crescent of the sun peeps out once more from behind the moon. The eclipse is over.

A total eclipse may last for as long as seven minutes, but the majority are far shorter. In spite of the invention of the corono-graph they are still of great importance for only at these times can the outer regions of the corona be seen or the stars directly ' beyond' the sun be observed.

Both Newton and Einstein forecast that the path of a ray of light would be slightly ' bent ' in the vicinity of matter. However, the amount of bending suggested by the two theories differs and it has long been hoped that by measuring the apparent shift in a star's position when its light has passed close to the sun at a total eclipse it would be possible to determine which theory is correct. Many eclipse expeditions have been organised with this as one of their prime objectives but so far, although the shift has certainly been observed, the accuracy obtained has not been sufficient to settle the argument!

Recent studies of the sun's corona have raised one of the most interesting problems of modern astronomy and one which underlines vividly how little we really know of the workings of even our nearest star. For many years the spectrum of the solar corona had revealed the presence of certain lines which could not be identified, and it was suggested that they might be due to an element unknown on earth, which was accordingly named 'Coronium'. In 1941 it was shown mathematically that these lines could be produced by the atoms of quite common elements such as iron and nickel if they were so highly ionised that they had lost as many as thirteen or more of their outer electrons. Such a loss could occur, however, only if the corona possessed a fantastically high temperature, between one and two million degrees.

Although astronomers had already become accustomed to the fact that the hotter chromosphere surrounded the cooler photosphere, they were hardly prepared to accept the fact that the corona was so extremely hot. The effect, studies have shown, arises from intense shock waves associated with solar flares or other activity originating in sunspots.

Enough has been said to show that the sun is an extremely interesting body to study, and one who guards his secrets well. The more we learn about him, the more there seems to discover. But new devices and new techniques are rapidly increasing our knowledge about the sun and solar activity. Orbiting solar observatories and rockets that go above the earth's atmosphere are giving information about solar radiations in the far ultraviolet and X-ray region, light that our own atmosphere heavily absorbs. And giant radio telescopes reveal other details about the active regions, the velocities of the clouds and their extent into space. Our space probes, like the one to Venus, transmit details of the outer fringes of the corona, which we have referred to as the solar wind. Undoubtedly the future holds many surprises in store.

What now of the sun's family?

THE SOLAR SYSTEM

The nine principal members of the sun's family are the planets Mercury, Venus, the Earth, Mars, Jupiter, Saturn, Uranus, Neptune and Pluto,.given in order outwards from the sun. Their main statistics have been tabulated in Appendix I. Between them they possess thirty-one known moons. In addition, over three thousand minor planets have been identified and no doubt there are thousands more which are as yet undiscovered, as well as countless millions of meteors and micro-meteorites and an un-numbered swarm of comets. Yet, formidable as this array may sound, the total matter possessed by all these bodies is estimated to add up to only about one-thousandth part of that in the sun.

The way in which these bodies move in their orbits was determined by Johannes Kepler early in the seventeenth century from his brilliant analysis of the observations of Tycho Brahe. This was one of the milestones in the history of astronomy, for Kepler's three laws have stood the test of time and have been found to apply not only within the solar system but throughout the entire universe. They are set out below with, in brackets, a simply worded explanation of what they really mean.

Kepler's laws stated : —

　1. The planets move in ellipses with the sun at one focus.
　2. The radius vector sweeps out equal areas in equal times. (Figure 8) (i.e. The closer a planet is to the sun, the faster it will move.)
　3. The square of the time of revolution is proportional to the cube of the mean distance. (i.e. There is a direct relationship between the average distance of a planet from the sun and the time it takes to complete one revolution.)

The orbits of the planets are shown in Figure 9. As the furthest planet, Pluto, is rather more than one hundred times as far from the sun as the nearest, Mercury, the two parts of the diagram have been drawn on different scales. Mercury takes

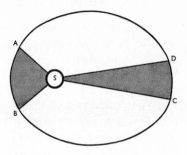

Fig. 8. Kepler's first law stated that a planet must move in an ellipse with the sun at one of the foci.

His second law stated that the radius vector would sweep out equal areas in equal times. Area S A B is equal to area S C D. As both are to be swept out in the same time it is clear that the planet must move much more quickly from A to B (when it is close to the sun) than it does from C to D (when it is furthest away).

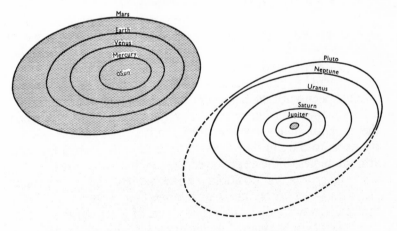

Fig. 9. The orbits of the planets drawn to scale. The orbit of Mars is shown by the small shaded area in the lower diagram. Note the marked eccentricity of the orbits of Mercury, Mars and Pluto.

The orbit of Pluto is inclined at an angle of about $15\frac{1}{2}°$ to that of Neptune, that part of the orbit which is 'below' the orbital plane of Neptune being shown dotted.

approximately 88 days to go round the sun and Pluto nearly 248 years, or about one thousand times as long.[1]

When describing the movements of the planets there are certain convenient terms in general use which it would be as well to understand. (See Figure 10.) A planet that is in a line with the sun and the earth is said to be in *conjunction*. Mercury and Venus being nearer to the sun than the earth are called the *inferior planets* and can of course, be in line either on the same side of the sun as the earth or on the opposite side. These two positions are therefore called *inferior conjunction* and *superior conjunction* respectively.

We have seen how flat the solar system is and how the orbits of the planets lie very nearly in the same plane as that of the earth. From the earth we therefore see the orbit of an inferior planet almost edge-wise on, as if we were looking at the edge of a plate with the planet travelling around its rim. In consequence, both Mercury and Venus appear from the earth to move from side to side of the sun and thus may sometimes rise ahead of him in the early morning and sometimes be left behind when he sets in the evening. Hence we often refer to them as Morning or Evening Stars. When the planets are at their greatest angular distance from the sun they are said to be at greatest *elongation*, either easterly or westerly. Venus can be as much as 47° from the sun, and on these occasions, owing to her great brilliance, can often be seen with the naked eye during the daytime.

The planets further from the sun than the earth are called the *superior* planets. When they are on the opposite side of the earth to the sun they are said to be in *opposition*, and when at right-angles to the direction of the sun, in *quadrature*.

By recording the positions of the planets and noting their periods of revolution, one can arrange them in sequence from the sun and determine the proper sizes of their various orbits. The ancient Greeks knew this and even the Babylonians drew the Solar System to scale. But they did not know the distance of any planet. If only they could find the distance to any one of the

---

[1] Those interested in simple mathematics might like to note the verification of Kepler's third law. Treating the figures for Mercury as unity, we find that for Pluto $100^3 = 1000^2$.

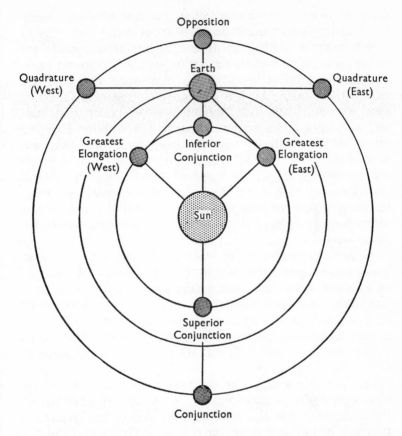

Fig. 10. Orbit of the earth, with one inferior and one superior planet.

planets, they could then determine the scale of their diagram and calculate all of the other distances.

We saw in Chapter II how astronomers calculated the distance of the moon by determining its position from two widely separated places on earth. The same procedure was used in 1675 with the planet Mars which was observed simultaneously from Paris and from Cayenne, in South America, and from these results Cassini calculated the distance of the earth from the sun as 86,000,000 miles, which was at least of

the right order of magnitude. Since that time numerous deter-
minations have been made, often using one of the minor planets
which, because of their small size, appear as points of light
rather than as discs as do the major planets and so are easier
to observe with accuracy. Some also come comparatively
close to the earth. Typical of these was the observation of
Eros, a strange little flying mountain some sixteen miles long
and only four miles across, which passed within 16 million
miles of the earth in 1931. A world-wide operation was
organised by Sir Harold Spencer Jones, then His Majesty's
Astronomer Royal at the Cape of Good Hope, to observe
this close approach. More than 2,000 photographs were taken
in scores of observatories throughout the world and the figure
calculated, 93,005,000 miles, received general acceptance for
some years.

As in the case of the moon, however, this figure has now been
verified and slightly modified by the use of radar. At inferior
conjunction Venus can come within about 25 million miles of
the earth, and her distance has been directly measured by
radio astronomers in England, the United States and Soviet
Russia. As a result, the most likely value is now believed to be
92,925,100 miles, although still with a possibility of error of
up to 30,000 miles.

The mean distance of the earth from the sun is of tremendous significance in astronomy, not only for fixing the size of
the solar system but also because, as we shall see in Chapter IV,
it is our jumping-off point for obtaining the distances of the
stars. This measurement has therefore been honoured with the
dignified title of *The Astronomical Unit*.

To the laymen the term ' astronomical figures ' normally
conjures up a picture of rows and rows of zeroes. Frequently,
however, astronomers prefer to deal, not with the enormous distances as such, but with the very minute angles which must
first be measured before these distances can be calculated. For
this reason it is more usual for them to talk about the *solar
parallax* rather than the sun's distance. The solar parallax is the
angle at the sun's centre that would be subtended by the
equatorial radius of the earth when at its mean distance, and

before the advent of radar this was accepted as 8·7948 seconds[1] of arc. Direct measurements of the distance of Venus by radar have now put this figure at 8·7943 seconds, a difference of only a few ten-thousandths of a second of arc but nevertheless representing a discrepancy in the position of even the nearer planets which might be as great as 10,000 miles. To a space probe (or ultimately a space ship) such an error in the position of its destination might be very inconvenient!

Before we leave the question of distance there is a most intriguing law connecting the spacing of the planets which, although it was formulated in 1772, has still not been satisfactorily explained. This is *Bode's Law*. To obtain it, write down the numbers 0, 3, 6, 12, 24 . . . as a series, adding 4 to each number and then dividing the answer by 10. If we call the Astronomical Unit ' One ' the distances of the other planets follow the sequence remarkably closely, as is shown in the following table : —

	Bode's Law	Actual Distance (in Astronomical Units)
Mercury	0·4	0·39
Venus	0·7	0·72
Earth	1·0	1·0
Mars	1·6	1·52
Asteroids	2·8	2·65 average
Jupiter	5·2	5·2
Saturn	10·0	9·54
Uranus	19·6	19·2
Neptune	38·8	30·1
Pluto	77·2	39·5

Of course Bode knew nothing of the planets beyond Saturn, which had not then been discovered; neither did he know of the existence of the asteroids. When these became known they fitted into their places almost exactly, with the exceptions of Neptune and Pluto. Astronomers, however, have found no physical explanation for the law, so that it is, in all probability, only mere coincidence.

---

[1] 1 degree = 60 minutes (of arc), 1 minute = 60 seconds (of arc), 1 second thus equals 1/3600 of a degree.

Let us now consider the planets individually in a little more detail.

MERCURY, THE ELUSIVE TWINKLER

Mercury, the smallest planet and the one closest to the sun, is often thought of as an elusive little fellow, difficult for the casual observer to pick out. Certainly it is necessary to choose the right evening when Mercury is near elongation, but on occasions he can appear as a brilliant twinkling star low down near the horizon.[1] Under favourable circumstances he may be visible for as long as an hour at dawn or dusk.

To the ancient Greeks, Mercury was the swift messenger of the Gods. We are told that in his youth he was also extremely mischievous but always managed to conceal his movements and so avoid being found out. This name is appropriate for this fast-moving little planet that so often evades detection!

Mercury can never appear more than 28° from the sun even at elongation and owing to his very eccentric orbit his greatest distance may sometimes be as little as 18°. Good conditions are therefore necessary if we are to see him without the help of a telescope. The best times occur in the spring shortly after sunset and in the early morning during the autumn, but clear skies near the horizon are often hard to come by, and many an amateur observer has never seen Mercury. (They are in good company for we are told that Copernicus never succeeded in seeing it.)

From the evidence of visual observations, astronomers concluded that Mercury rotated on its axis in 88 days, the same time that it made a single revolution around the sun, thus always exposing the same face to the sun's scorching rays.

Radio astronomy, however, gives a different and apparently conclusive answer. Using the great 1,000-foot dish, scientists at the Arecibo Ionospheric Observatory in Puerto Rico sent radar pulses to the Planet Mercury and detected the return signal. If the surface of Mercury were perfectly smooth they would get no information about its rotation from the radar echoes. But reflections from mountains or other irregularities produce

[1] At one time Mercury was known to the Greeks as *Stillbon*, ' the twinkler '.

changes in frequency of the reflected signal, by the so-called Doppler effect. When planetary rotation causes the mountain to be moving away from the earth the frequency of the return signal is less than that of the transmitted; when the mountain is approaching the earth the frequency is greater. A detailed analysis of these signals clearly indicates that the planet Mercury takes about fifty-nine days to complete one rotation. Also, the direction of rotation is similar to that of the earth.

Mercury's distance from the sun varies between 29,000,000 and 43,000,000 miles, the mean being about 36,000,000 or a little more than a third of the earth's distance. Every square mile on the sunny surface of the planet thus receives seven times as much heat as a corresponding square mile on the earth. Apart from the sun, the sunny side of Mercury must therefore be the hottest place in the solar system. The temperature is so high that a lump of lead would quickly melt in the sunshine. No normal liquids could exist, for they would all boil away.

No atmosphere has been detected on Mercury and we can be fairly certain that none exists. Because of its low mass, the planet has a weak gravitational pull so that the velocity of escape is small, only 2·6 miles per second. Any gases originally present would long since have escaped.

In size and general appearance Mercury is very similar to the moon. Its diameter of 3,100 miles is only slightly greater. The surface of Mercury appears to be covered with the same type of brownish volcanic ash, making it an arid, desolate, dust bowl of sunlight. The planet's albedo[1] is only 7 per cent, also about the same as the moon.

As noted above, Mercury takes a little less than 88 earthly days to travel around the sun, and thus completes the trip four times in one of our years. Because of the slow rotation period, the dark side of the planet has an excellent chance to cool off. However, it seems certain that no life could possibly exist on Mercury.

'VULCAN'

Newton's theory of gravitation predicted that the orbits of the planets would not remain fixed in space but that their major (longer) axes would revolve very slowly. This was observed to

[1] See Page 40.

occur but in the case of Mercury the movement was appreciably greater than was predicted, and for a long time astronomers thought that there must be another planet even closer to the sun than Mercury, whose 'perturbations' would account for the difference. Although this planet could not be found they became so sure of its existence that it was actually named 'Vulcan'.

One of the triumphs of Einstein's theory of relativity is that this apparent discrepancy in the rate of rotation of Mercury's orbit has been accounted for, and no further 'planet' is required.

VENUS, THE PLANET OF MYSTERY

If Mercury is elusive, Venus, the next planet, is frequently the most prominent object in the sky. All of us at some time must have noticed her hanging in the west like a brilliant white lamp in the evening twilight, at times not setting for more than four hours after the sun. Yet for all her brilliance Venus is a planet of great mystery who, unlike her namesake in mythology, tries hard to keep her secrets well hidden.

Venus takes approximately 225 days, seven and a half months, to go once round the sun, moving in an almost circular orbit at a distance of about 67 million miles. She is very similar in size to the earth, with a diameter only about 200 miles shorter, and is often referred to as our twin sister. In other ways, however, she is very different. Although she can sometimes approach to within about 25 million miles of the earth, closer than any other major planet, we have never been able to see her surface because it is perpetually hidden in a thick mantle of cloud. Clouds are excellent reflectors of sunlight and this is the reason for her brilliance, the albedo of Venus being about 59 per cent. So far, however, no form of light in which we can photograph Venus has penetrated her mantle, and it is an intriguing thought that if there are any conscious living creatures on this twin of ours, they have probably never seen the stars, and may well be unaware of the existence of the rest of the universe — with the probable exception of the sun.

Studying the atmosphere of another planet presents many problems. The stars, as we know, are suns, shining with their

(a)

(b)

(c)

1. The 'Waxing' Moon. The Moon aged (a) 4 days, (b) 7 days (1st quarter) and (c) 10 days, showing the movement of the terminator. Craters stand out in sharp relief shortly after 'sunrise' on the moon and many tall peaks are lit up while still within the dark zone. (Lick Observatory photograph.)

2. The Full Moon. Many of the conspicuous features in Plates 1 and 3 have become practically invisible due to the high-lighting of the surface, but many ray systems have become prominent. Compare slight differences in the position of 'Sea of Crises' on Plates 1 and 2, due to libration. (Lick Observatory photograph.)

1. Sea of Clouds.
2. Sea of Moisture.
3. Ocean of Storms.
4. Sea of Rains.
5. Sea of Cold.
6. Sea of Serenity.

7. Sea of Tranquillity.
8. Sea of Fertility.
9. Sea of Nectar.
10. Sea of Crises.
11. Marginal Sea.
12. Smyth's Sea.

Craters: 13. Plato.
14. Archimedes.
15. Copernicus.
16. Alphonsus.
17. Tycho.
18. Grimaldi.

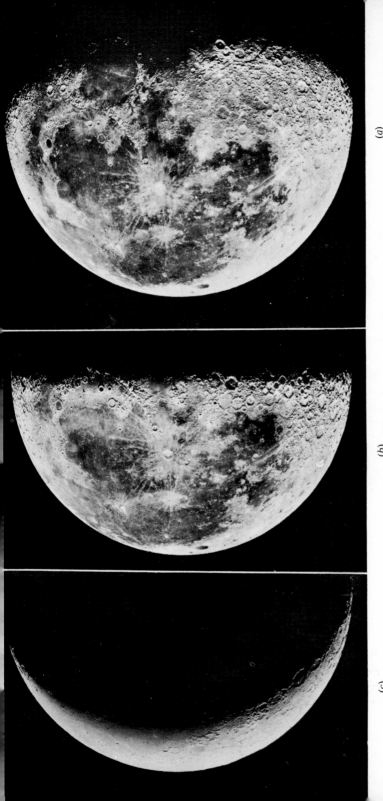

(a)

(b)

(c)

3. The 'Waning' Moon. The Moon aged (a) 20 days, (b) 22 days (last quarter), and (c) 26 days. The giant crater Schickard is conspicuous in the lower part of (c). (Lick Observatory photograph.)

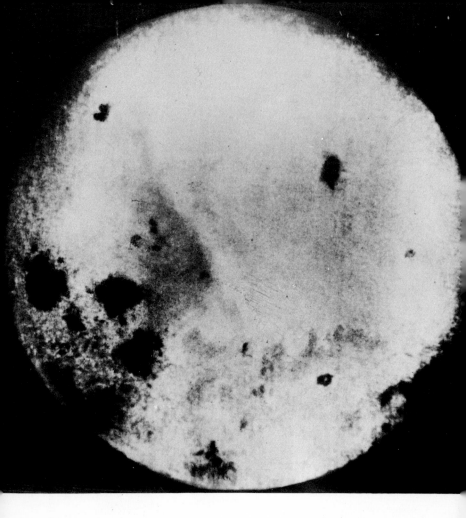

4. The 'Back of the Moon'. Features to the left of the photograph belong to 'our' side of the moon and can be identified from Plate 2. The Sea of Crises is conspicuous with, below it, the Sea of Fertility. The two large dark areas close together are Smyth's Sea and the Marginal Sea which appear to us on the western limb. Note the scarcity of 'seas' compared with Plate 2. (*Soviet Weekly* photograph.)

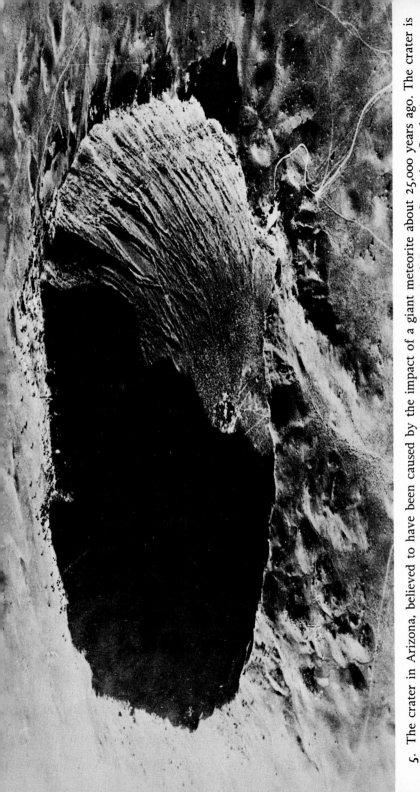

5. The crater in Arizona, believed to have been caused by the impact of a giant meteorite about 25,000 years ago. The crater is 600 feet deep and measures almost a mile across. (Photograph: United States Information Service.)

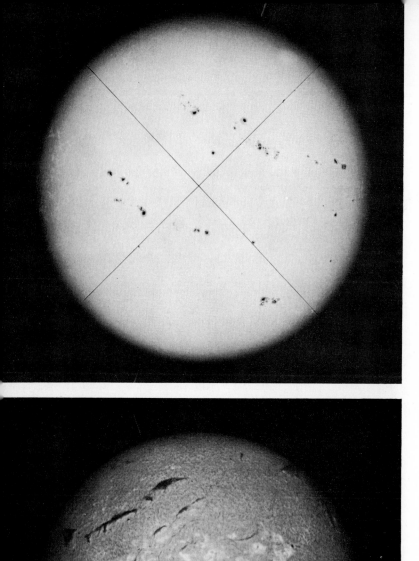

6 (top left). The sun's disc photographed on 24 May 1947, about the time of sunspot maximum. Notice the lines of sunspots on either side of but not on the equator. The sunspot near the centre of the photograph has a diameter roughly equal to that of the earth. (Royal Observatory Greenwich photograph.)

7 (bottom left). The sun photographed in the light of Hydrogen (Hα). Note the granulations and the bright faculæ, some of which are surrounding sunspots; also the dark plumes which are solar prominences situated high above the photosphere.

8. The total eclipse of 9 May 1929, showing the solar corona and, on left-edge of the sun, a large prominence. (Photograph: J. Miller, Sproul Expedition.)

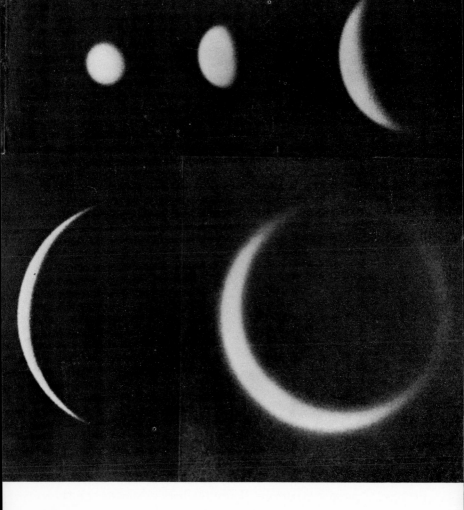

9. Venus photographed at different phases but with constant magnification, showing the changes in apparent size. The lower right-hand photograph was taken near inferior conjunction, when Venus passed between the earth and the sun, and shows the effect of the bending of the sun's rays by the atmosphere of Venus. (Photograph by E. C. Slipher, Lowell Observatory.)

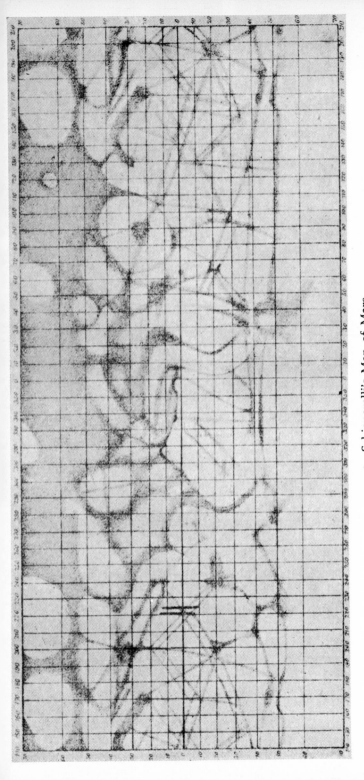

10. Schiaparelli's Map of Mars.

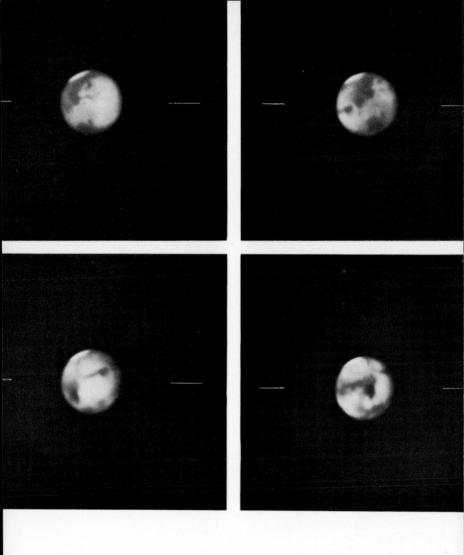

11. Four views of Mars showing the changing surface features as the planet rotates. A polar cap is very prominent. (Photograph by W. S. Finsen, Johannesburg Observatory.)

12. Two photographs of Jupiter showing changes in the cloudy bands on his surface. The Great Red Spot is well seen in the top photograph. In the bottom photograph it has almost disappeared, whilst Ganymede, one of Jupiter's moons, is seen, with its shadow cast on the body of the planet. (Palomar Observatory photograph.)

13. Saturn, showing the ring system. The 'A' ring, the bright 'B' ring and Cassini's Division are clearly seen; also the cloud bands on the body of the planet, and its shadow on the rings. (Mount Wilson Observatory photograph.)

14. Three views of Halley's comet taken at its apparition in 1910, showing the way in which the visible tail varies. The star trails are caused by the length of the exposure. The bright object in the central photograph is the planet Venus. (Photographs by Knox-Shaw, Helwan Observatory and E. C. Slipher, Lowell Observatory.)

15. Part of the rich Star Cloud in *Sagittarius*, believed to be in the direction of the centre of our galaxy. (Photograph: E. E. Barnard, Mount Wilson Observatory.)

16. The Small Cloud of Magellan, visible from the southern hemisphere. Probably a satellite of our galaxy. (Harvard Observatory, Arequipa.)

17. The Crab Nebula in Taurus (M.1). The debris of the supernova observed in A.D. 1054, which is still expanding. (Palomar Observatory photograph.)

18. The Pleiades. The nebulosity consists of interstellar dust, beautifully illuminated by starlight. (Photograph: Roberts, Crowborough.)

19. The Globular cluster M13, seen resolved into stars, of which some fifty thousand have been counted. The condensed central region is about 10 light-years across, and the distance of the cluster is of the order of 25,000 light-years. It is just visible to the naked eye in the constellation of Hercules. (Palomar Observatory photograph.)

20. One of the largest planetary nebulae, M97, in Ursa Major. Known as the 'Owl' Nebula. The very hot central star excites the extremely rarefied surrounding nebulosity causing it to glow. (Mount Wilson Observatory photograph.)

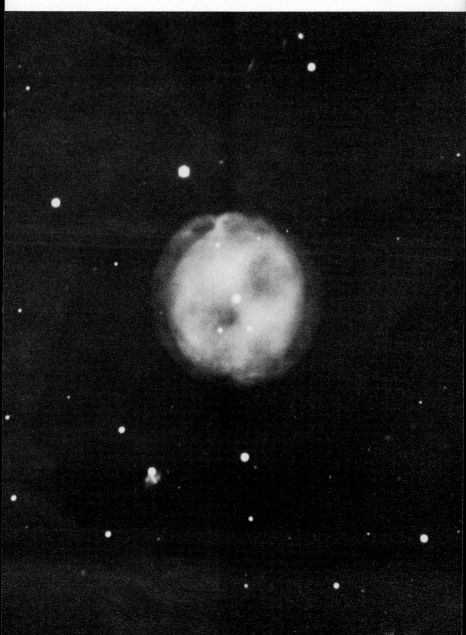

21 (above). The Great Nebula in Orion (M42). Without doubt one of the finest of all the nebulae. Visible to the naked eye as the middle star of Orion's sword. See Plate 22. (Photograph: Richey, Mount Wilson.)

22 (top right). The constellation of Orion showing rich nebulosities. The three stars of the 'belt' are in the middle of the photograph with the hazy 'sword' stars below it. (Photograph by Ross, Yerkes Observatory.)

23 (bottom right). Part of the Milky Way, showing the Southern Cross and the 'Coal Sack', a fine 'dark' nebula. Two of the stars of the Southern Cross are red and so do not show up well in the photograph. (Photograph by Franklin-Adams, Johannesburg Observatory.)

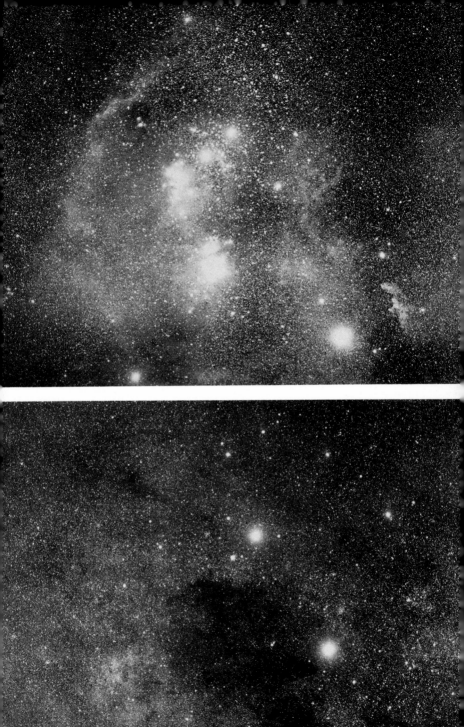

24. The 'Horsehead' nebula in the constellation of Orion to the south of Zeta Orionis. An excellent example of dark dust showing up against bright gaseous nebulosity. (Palomar Observatory photograph.)

25. The Great Spiral Galaxy in Andromeda (M31). One of the nearest 'spirals', which is probably similar to but slightly larger than our own galaxy. It is easily visible with the naked eye under good conditions. The photograph shows a bright central nucleus (Population II stars) and extensive spiral arms containing bright new (Population I) stars and much nebulosity. Two elliptical galaxies, satellites of the main spiral, are clearly seen. (Palomar Observatory photograph.)

26. Spiral Galaxy (M33) in Triangulum. The spiral structure and trailing arms are clearly seen, although this galaxy does not appear to be as far developed as M31 (Plate 25). (Mount Wilson Observatory photograph.)

27. The Spiral Galaxy (M51) in Canes Venatici. A fine spiral seen
almost 'full face'. The 'catherine wheel' rotational appearance is very
marked. (Palomar Observatory photograph.)

28. The Spiral Galaxy (N.G.C. 4565) in Coma Berenices. This spiral is seen 'edgewise on', showing the central hub and the dark absorption effect of the gas and dust in the central plane of the galaxy.
(Palomar Observatory photograph.)

own light and so sending us messages about themselves which can be decoded with the aid of a spectroscope. A planet, on the other hand, can only be studied by means of the sunlight it reflects. This light is affected not only by the atmosphere of the planet but also by its passage through the atmosphere of our earth, and the two effects must of course be disentangled if we are to learn anything about the planet itself. One way of doing this is to compare the light reflected by the atmosphere of a planet with that reflected by the moon, which has no atmosphere. The two will be found to differ and it may thus be possible to identify which absorptions are due solely to the atmosphere of the planet, and so learn of what it is made. Unfortunately this interesting piece of detective work encounters certain difficulties. If the constituents we are looking for in the planet are also abundant in our own atmosphere they may not show up at all. Oxygen and water vapour clearly fall into this category. Again, if the absorption lines of an element lie mainly in those parts of the spectrum which are cut off by our atmosphere, we obviously cannot learn about it by this method.

These difficulties have to some extent been overcome by other lines of approach, but modern astronomy is now tackling this problem in two new and very ambitious ways. In the case of the nearer planets, deep-space probes are being sent to take a close look at our neighbours, with instructions to send back messages describing what they find. This method is already yielding fruitful results, and much has been learnt from these latest ambassadors, not only of conditions on the planets they have visited but also of those in the interplanetary space through which they have travelled. At the same time ' observatories ' sent up by balloon or by artificial satellite are being used to investigate the other bodies in the solar system (as well as much of the universe as a whole) from positions above the most troublesome layers of the earth's atmosphere. Balloons are obviously very limited in the height they can attain, although this method has been used with great success to obtain photographs in ultra-violet light from twenty-three miles up, and above the ozone layer. Satellites, on the other hand, operate very much higher and, being above the ionosphere, can study the radiation received over the whole electro-magnetic spectrum. In this way a whole new field of investigation is gradually being opened up as new 'observatories' are put into space.

With the help of these new methods, it has become possible to obtain a fairly comprehensive picture of the atmosphere of Venus. It has long been known that carbon dioxide is present in quantities many hundreds of times more plentiful than on earth, but water vapour, which could only be detected with difficulty from our terrestrial observatories, has also been detected. Oxygen, on the other hand, appears to be almost completely absent. Clearly, from a study of the atmosphere alone, it appears highly improbable that intelligent life can exist on this neighbour of ours.

The surface temperature of Venus is unpleasantly high by earthly standards. The American space-craft Mariner II which made the first successful fly-by of the planet on 14th December 1962, passed within 21,000 miles of Venus after a journey of more than 220 million miles. Measurements obtained from this 'probe' show an average surface temperature in excess of 400°C., considerably higher than expected. The upper layers of the Venusian clouds are cold, with a temperature of some −40°C.

The clouded surface of Venus shows vague shadows from time to time, but such markings are not permanent, and one cannot determine from them the true period of planetary rotation. The best observations, by means of radar technique, indicate that the length of the Venusian day is about 250 days, with an uncertainty of some five days. More important, this figure indicates that the direction of rotation is retrograde, in the opposite sense of that in which most of the planets of the solar system rotate.

An observer on Venus, if he could see the sun at all, would notice that it rose in the west and set in the east, taking about half a year to go from horizon to horizon.

Any owner of a small telescope can observe the changing phases of Venus as she circuits the sun. When she is on the far side of the sun from the earth, we see a 'full' Venus, but because she is then at her greatest distance from us, about 160 million miles, she is not as bright as when we see her as a crescent, very much closer to the earth. The change in the apparent size of Venus is clearly shown in Plate 9, from which it can be seen that she appears at her brightest (i.e. has the largest apparent illuminated area) shortly before or shortly after inferior conjunction. Although at this time less than a quarter of the visible surface is illuminated, she may reach a stellar magnitude of −4·4.[1]

In other ways our sister planet is a most tantalising object for the observer. Markings can undoubtedly be seen with telescopes of some power, but they are transient and never well defined. Shady areas or streakiness may appear one night and have entirely disappeared the next. On occasions considerable brightening may be observed near the ends of the horns of the crescent, known as the cusps, or the cusps themselves may extend a long way round the ' dark ' side of the planet, owing to the illumination of its atmosphere. They may even appear to form a complete ring. At other times the ' dark ' part of Venus may be seen shining dimly in much the same way as does the crescent moon, although we can be sure that owing to the distance of Venus this could not possibly be due to ' earthshine '. This effect is called the *ashen light* and has often puzzled observers. Although we cannot be absolutely sure of its origin, astronomers have suggested that the ashen light may arise as a sort of aurora borealis in the upper atmosphere of Venus, luminosity excited by particles from the sun.

Although this mysterious planet still holds many secrets we can be sure that the time cannot be far off when much of our curiosity will be satisfied.

MARS: ANOTHER ABODE OF LIFE?

Our next-door neighbour as we move outwards from the sun is the fiery-red planet Mars. To the Romans Mars was the God of the year, particularly of the spring-time, but he was later identified

----

[1] See Page 131.

with Ares, the Greek God of War, who loved fighting for its own sake. Even up to comparatively recent times each opposition of Mars was regarded with foreboding in certain countries.

To-day this is perhaps the most interesting of all the planets to us on earth, as with the prospect of inter-planetary travel firmly established we ask ourselves the question ' Is there really life on Mars? ' or perhaps even ' Will man find conditions agreeable when he gets there? '. At the turn of the last century the existence of intelligent beings on Mars was confidently assumed in many (though not all) astronomical circles. How flimsy was the evidence for this belief we shall very soon see.

So far, as we have moved away from the sun, each successive planet has been slightly larger than the one before, but Mars is the exception, having a diameter of only 4,216 miles, or a little more than half that of the earth. Mars rotates on his axis once in 24 hours 37 minutes, so that day and night are very similar in length to our own. A year, however, lasts for 687 earthly days (or 669 Martian days), and as the axis of Mars is inclined at about 25° to the plane of its orbit, the planet undergoes a succession of seasons, each of which will be almost twice the length of ours.

Mars moves in a very eccentric orbit with a mean distance from the sun of about 141½ million miles. The actual distance varies between 129 and 154 million miles which, when we remember Kepler's second law, tells us that his speed in his orbit will vary very considerably, and so, therefore, will the length of the seasons. When Mars is near perihelion and moving most rapidly, his southern hemisphere is enjoying spring and Summer. These seasons, however, last only 306 days whereas the autumn and winter which occur when Mars is near aphelion, and so moving more slowly, last 381 days. In the northern hemisphere the seasons are, of course, reversed. (On earth the similar discrepancy is only seven days, summer in the northern hemisphere being the longer period.)

Approximately every 780 days the earth overtakes Mars and races past him, having the advantage of the ' inside of the bend '. Mars then makes one of his periodic close approaches to the earth, as he passes through opposition, and is well placed for observation. Under favourable circumstances he can approach within about 35 million miles of the earth but on other occasions, owing to his very eccentric orbit, he may not come closer than about 63 million. This is clearly shown in Figure 11, from which it can be seen that 1956 was a very close opposition and that the next really favourable one will not occur until 1971. It was most unfortunate that in 1956 what appeared to be dust clouds in the atmosphere of Mars prevented astronomers from

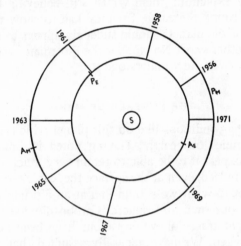

Fig. 11. The distance of Mars from the earth at opposition may vary between about 35½ and 62 million miles. The aphelion and perihelion points on their orbits are shown. 1956 was a favourable opposition, but the next favourable one will not occur until 1971.

seeing much of his surface at a time when he was otherwise very well placed for observation.

We have all at some time been in an express train when it is overtaking and racing past a slower train, and we will have noticed how the slower train appears to be moving the 'wrong way' against the background. The earth is travelling at some 18½ miles per second in its orbit round the sun. As it passes Mars, doing a mere 15 miles per second, and in an outer lane, the same effect is seen and Mars *appears* temporarily to move the 'wrong way' against the background of stars. Once the earth is past and clear, however, Mars is again seen to resume his normal passage from west to east amongst the stars. At these times he is said to describe a 'loop' (Figure 12) and each of the superior planets can be seen doing the same thing at about the time of oposition. Such loops were, of course, known to the early astronomers as they carefully followed the progress of the planets in the zodiac. We can well understand the difficulty they had in explaining them whilst still believing that everything went round the earth, for they had to show the reasons why some of the planets should suddenly appear to turn round and go the other way! No wonder the Ptolemaic system was so complex that very few understood it.

THE SURFACE OF MARS

What are the conditions like on this planet which, in so many ways, resembles our earth? Having noted its small size we should not expect it to be able to retain very much of an atmosphere, and indeed, even at the surface the 'air' is so thin that it is equivalent to what we should find at an altitude from 15 to 20 miles in our own atmosphere. Carbon dioxide is certainly present. Slight traces of water vapour have been detected but none of oxygen. We have generally assumed that nitrogen is also present, perhaps in some abundance. However, there is no clear-cut evidence proving either the presence or the quantity of this substance. Argon, a slightly heavier gas, may also be present.

Photographs of Mars taken by the Probe Mariner IV and transmitted to earth indicate that the planet closely resembles our moon. Thousands of craters, ranging in size from 100 miles or so down to 2 or 3 miles, pockmark the surface. The peaks of some of these craters appear to be covered with hoar-frost. The pressure of the atmosphere at the surface of Mars appears to be not greater than 2 or 3 per cent of the air pressure at sea level on earth.

The Greek name for Mars was *Ares*, and in our next chapter we shall be meeting a brilliant red giant star *Antares*, whose name means 'The rival to Ares'. Mars really does appear red in the sky because he has a predominantly red surface, believed to be due to the presence of large quantities of iron oxide (iron rust) and this, in turn, may account for the paucity of oxygen in the atmosphere. It seems probable that much of the surface will resemble a barren, rusty desert without a trace of moisture.

What then of the famous canals? At the opposition of 1877, which was a particularly favourable one, the Italian astronomer Schiaparelli saw and mapped a number of markings on the sur-

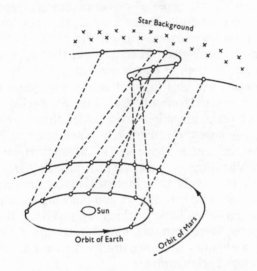

Fig. 12. This diagram shows the reason for the apparent 'retrograde' motion of a superior planet against the stellar background as the earth over-takes and passes him.

face which he described as 'canali', a word which in his language meant 'channels' (Plate 10). It is easy to see how reports of his findings could be misinterpreted to indicate 'canals'. Schiaparelli himself never seems to have stated that he thought that the markings were artificial, but this did not deter others with more vivid imaginations. Fantastic stories were circulating towards the turn of the nineteenth century about a beleaguered Martian population who had constructed an extensive network of canals in order to preserve and make the best use of their woefully short water supply. The name of

America's great astronomer Percival Lowell, in particular, is associated with Mars, and he made very extensive studies of the planet, mapping over 400 'canals' and writing exhaustively on the idea of a Martian population.

On the other hand there have been many very competent observers who would not admit to having seen the 'canals' at all, although markings of some sort undoubtedly do exist and can be seen with telescopes of moderate power. (See Plate 11.) They have been carefully mapped and all the principal features have now been given names, as on the moon. Generally the markings are greyish-green in colour, but they appear to show definite variations both in colour and extent with the changing Martian seasons and this first led to the suggestion that they might be due to some form of vegetation. We are led to wonder what sort of plant life might exist under these very rigorous conditions, for near the equator the temperature at mid-day is unlikely to rise much above 50°F., a mild winter's day in America, whilst during a polar night it probably falls to about −90°F. This is roughly the range of temperature encountered on earth by Antarctic expeditions, and there are certainly plants of the moss and lichen variety which can survive in those inhospitable regions. More than this, however, we are unlikely to be able to say with any confidence until someone actually goes to Mars and brings back a sample!

At either pole of Mars white caps can be discerned gleaming in the sun. Owing to the shortage of water these cannot be thick ice caps, as are found on earth, but are believed to consist of layers of 'snow' or possibly some form of hoar-frost, and are unlikely to be more than a few inches thick. As they melt with the onset of summer, they can be seen receding and this change appears to be connected with the spread of the darker areas. It seems possible that it indicates natural irrigation although alternative explanations have been proposed and we would be unwise to jump to any hasty conclusions.[1]

We cannot leave Mars without taking a look at his two ridiculous little moons, Phobos and Deimos (the Greek names for fear and terror, the attendants of Mars), which are believed to be only about ten miles and five miles in diameter respectively.

Because of the normal brilliance of Mars and their close atten-
dance on their parent they were not discovered until the favourable opposition of 1877, although their presence had been
remarkably foretold 151 years earlier by Dean Swift in his epic
Gulliver's Travels. Writing of the Laputan astronomers he
said : —

> 'They have likewise discovered two lesser stars or satellites which
> revolve about Mars, whereof the innermost is distant from the
> centre of the primary planet exactly three of his diameters and
> the outermost five. The former revolves in the space of 10 hours,
> the latter in $21\frac{1}{2}$.'

This was incredible guesswork. In fact Phobos revolves at a
distance of roughly $1\frac{1}{2}$ diameters and Deimos at $3\frac{1}{2}$. Their
periods of revolution are 7 hours 39 minutes and 30 hours 18
minutes respectively, and these factors introduce some interesting and rather unusual results. Deimos goes round Mars four
times whilst Mars rotates five times on his axis. Deimos thus
drifts very slowly across the sky, taking about two and a half
days to go from horizon to horizon. Phobos, on the other hand
goes round so fast that it 'overtakes' the rotation of Mars, and
so rises in the west and sets in the east. It goes three times
round the sky each day, so that in the case of Phobos the Martian
day is three times longer than its lunar month! Perhaps these
little fellows will make useful space stations for the first Martian expedition, but travellers will need to be careful as the
force of gravity will be negligible. Any violent movement might
send the astronaut soaring off into space again.

JUPITER—THE GIANT

The first four planets we have met have been solid little lumps
of rock similar to the earth. Mercury, Venus, the earth and

---

[1] Pictures transmitted from Mariner IV, which made a remarkable close
'fly-by' Mars in July 1965, passing only 7,000 miles from the planet, have
shown that there is a marked similarity between the features there and those
on the moon. Craters varying in size from 3 to 75 miles across can be distinguished, some near the South pole appearing to be ringed with frost. The
'canal' features do not seem to be prominent from close to.

Mars are, in fact, often referred to as the terrestrial planets. The next four, however, are very different, and would be best described as the giant planets. The first of them, and by far the largest, is Jupiter, who is not only the greatest planet in the solar system, but also contains more matter than all the others put together. Jupiter is so huge that more than thirteen hundred earths could be packed inside him and there would still be room to spare. Even so, it would take a thousand Jupiters to make a ball the size of the sun.

If Mars is sometimes rather disappointing to the amateur observer, Jupiter is probably the most rewarding of all. His distance from the sun is more than five times that of the earth, yet although he seldom approaches us to nearer than 400 million miles, even the simplest form of optical aid is sufficient to show him as a disc. This great giant has a diameter at his equator of 88,700 miles, more than eleven times that of the earth, but owing to his very rapid rotation the distance from pole to pole is only 82,800 miles, or some 6,000 miles shorter. A point on his equator takes only 9 hours 50·5 minutes to complete a revolution and must be moving at the amazing speed of 28,000 m.p.h. Nearer the poles the rate is very slightly slower (9 hours 55·7 minutes) which, as in the case of the sun, indicates that the body has an extensive atmosphere. Fortunately the immense gravitational pull of Jupiter amply makes up for the centrifugal force caused by his rapid rotation, so that objects do not tend to be thrown off into space. His surface gravity is some 2·6 times that of the earth, and an average sized man landing there would suddenly find himself weighing more than 1,200 pounds.

Jupiter's journey round the sun takes just under twelve earthly years and so we see him spending roughly a year in each of the twelve constellations of the zodiac. As his axis of rotation is very nearly 'upright' with respect to the plane of his orbit, he will not, of course, experience any changes of season.

A telescope of moderate power will enable the observer clearly to distinguish the dark reddish-brown bands on the disc but, as in the case of Venus, we are only looking at the top of the atmosphere which is believed to be some 8-9,000 miles in depth. (Plate 12.) This gaseous mantle appears to consist largely of hydrogen with some helium and with the addition of small quantities of methane (marsh gas) and ammonia. It would certainly be remarkably unpleasant to breathe. Owing to the planet's rapid rotation his temperature remains fairly constant at about −130°C., which is much the same as we found at 'midnight' on the moon. This is a temperature at which we should expect ammonia to be solid, and it is probable that many of the clouds we see are, in fact, clouds of frozen ammonia crystals. With such an atmosphere Jupiter is a brilliant reflector and with an albedo of 44 per cent he is second only to Venus.

What lies beneath this great mantle must, of course, be largely conjecture. Despite his immense size, Jupiter only 'weighs' about 317 times as much as the earth, so he cannot be more than about a quarter as dense. This has led to the suggestion that he probably has a solid metallic-rocky core not more than about 38,000 miles across, encased in a layer of 'ice' perhaps as much as 17,000 miles thick, the whole being enveloped in a dense, poisonous atmosphere roughly 9,000 miles in depth. Certainly such a world seems unlikely to harbour any living organism.

Of great interest to observers of Jupiter is the famous 'red spot' (Plate 12), situated just south of his equatorial belt, which has been seen on and off for more than a century. In size it is some 30,000 miles long by 7,000 miles wide, covering an area roughly equivalent to the whole surface of the earth; at times it is very prominent whilst at others it practically disappears altogether. Its colour, too, changes from deep red to pale pink or almost white. Although there is as yet no generally accepted theory as to what it is, the 'spot' is clearly 'floating' in the atmosphere, and may well consist of some form of metallic vapours thrown up as the result of a tremendous disturbance, possibly volcanic, deep down in the solid rocky core of the planet.

Recent radio observations of Jupiter have added still further to his mysteries by showing that he sometimes emits enormous bursts of energy from within his atmosphere. The origin of these outbursts is still obscure but if, as seems possible, they are due to 'thunder storms' of some kind, they must be millions of times more powerful than those on earth. There are also indications that Jupiter may have an intense magnetic field, so that he may well be surrounded by radiation belts of the type recently discovered by Van Allen around our own earth.[1] Certainly the more we study this planet the more forbidding he appears to be and any idea of visiting him becomes less and less inviting.

A pair of binoculars directed onto Jupiter may reveal his four brightest moons, which were first discovered by Galileo in 1610. These are always a delight to observe as they circle rapidly round their parent, going into eclipse behind him,[2] or passing in front of his disc, and constantly changing their relative positions. They have been named Io, Europa, Ganymede and Callisto in order outwards. Io and Europa are comparable in size with our own moon; Ganymede and Callisto, both being over 3,000 miles across, are almost the size of Mercury. It was not until 280 years after Galileo that another moon was discovered, Barnard finding a fifth in 1892, a tiny flying mountain only 100 miles across, closer to Jupiter than the other four. Seven more have been discovered since, the most recent in 1951, and all are so small that it seems probable that they were originally asteroids[3] which have been captured. Four of them revolve 'the wrong way' round Jupiter and their orbits are said to be retrograde; two are at distances of over fourteen million miles, almost half the distance of Mercury from the sun. As appropriate for the largest planet, he also has the largest family and holds sway over a tremendous region of the solar system, markedly perturbing the motions of every one of the other planets.

SATURN—AND HIS RINGS

Saturn is a planet of great beauty and without doubt the show-piece of the solar system. Viewed with a telescope the planet has a soft, rich, creamy appearance, while the rings can show

---

[1] Pages 50-51.

[2] It was a discrepancy in timing the re-appearance of Jupiter's moons when observed near opposition and near conjunction which led Roemer to realise in 1675 that light had a finite velocity. The discrepancy arose because of the time taken for the light to travel across the diameter of the earth's orbit, and from this Roemer first measured its speed.

[3] Page 116.

up in a number of pale pastel shades of blue and grey. The whole effect is a delight to behold.

Saturn is the furthest planet that was known to the ancients. Nearly twice as remote as Jupiter, he is about nine and a half times as far from the sun as is the earth, and moving in an eccentric orbit at an average distance of about 886 million miles, he takes just under 29½ years to complete a revolution.

As the second largest member of the sun's family, Saturn is in many ways similar to Jupiter, although somewhat smaller and about 20°C. colder. Like Jupiter he is in rapid rotation, a 'day' at his equator lasting only 10 hours 38 minutes. He is thus very oblate and when seen through a telescope has the shape of an orange, 75,100 miles across, nine and a half times the diameter of the earth,[1] but only 67,200 miles from pole to pole. Although occasional markings can be seen on his surface, including dusky belts round his middle and a periodical white spot, these are much fainter and more elusive than those on Jupiter. Otherwise his make-up is probably very similar although his atmosphere is more extensive and contains a higher proportion of methane. Once again, no form of life seems possible there. The density of Saturn has been found to be only about seven-tenths that of water, a most surprising result which means that if this planet could be immersed in water it would actually float!

It is the rings of Saturn that chiefly intrigue the amateur observer. With his optic tube Galileo saw them first as small bulges on the edges of the planet, but was disturbed to find the following year that they had disappeared. We understand now that as Saturn moves round the sun, the aspect of the rings as seen from the earth is constantly changing. The rings lie precisely in the plane of Saturn's equator, which is tilted at an angle of about 27° to the plane of his orbit. Sometimes we see them wide open with, say, the north pole of Saturn tilted towards the earth. Seven and a half years later (a quarter of the way round

---

[1] Saturn is roughly 9·5 times the earth's distance from the sun, and the earth's diameter. His mass is 95 times that of the earth, and he takes 29·5 years to go round the sun.

Saturn's orbit) they will be seen edge-wise on, and may be invisible except in the most powerful telescopes, while after a
further seven and a half years they will again be wide open
with Saturn's south pole tilted towards the earth. This cycle is
repeated in reverse as Saturn moves round the second half of
his orbit and so back to where he started, 29½ years later.

Plate 13 shows the rings of Saturn 'open', so that the many
components and divisions can be seen. These rings have a
tremendous span, measuring some 169,000 miles from edge to
edge (two-thirds of the distance from the earth to the moon),
yet are believed to be no more than 10 miles thick. Stars have
been observed shining right through them, and it will be recalled
how in our model in Chapter I we found that they could be
represented by a sheet of thin tissue paper six and a half inches
across.

The outer ring (called 'A' ring) is about 10,000 miles wide
and is generally a greyish-white in colour melting gradually
towards the edges. It is separated from the more brilliant 'B'
ring by an 1,800-mile gap known as Cassini's division, a gap
which goes right through the rings. Ring 'B' has a bright white
outer edge which fades evenly to a greyish interior, sometimes
tinged with blue, while inside these two is the shadowy, almost
slate-coloured crepe ring ('C' ring) which is much harder to
detect and was not discovered until many years after the other
two. This, too, is about 10,000 miles across, extending almost
to the body of the planet; there would just be room to fit the
earth into the gap between them. The shadow cast by Saturn
may be observed on the rings, as also can the shadow of the
rings themselves on the body of the planet. A telescope of
reasonable size (at least 5 inches) is, however, required to show
to the full the delicate shading and other delightful effects of
the rings, or the detailed markings on the surface of the globe
itself.

The rings of Saturn are highly reflective and the brightness
at which we see this planet depends far more on the aspect of
his rings than on his actual distance from the earth. Such a
wonderful broad expanse might at first sight appear to be an
ideal landing ground for a space expedition, but alas they would

be doomed to disappointment. It can be shown mathematically
that the rings could not exist at all if they were solid, as they
would quickly break up into fragments. Instead, they are
believed to consist of myriads of little particles, possibly lumps
of rock, each one a tiny moon travelling in a completely independent orbit in the plane of Saturn's equator. Just as we have
seen that the closer a planet is to the sun, the faster it must
move in its orbit, so it has been discovered with the help of the
spectroscope that the inner parts of the ring system are moving
much more rapidly than the outer parts, again showing that
the rings cannot be solid.

It is now believed that they were formed as the result of the
break-up of what may once have been an ordinary moon of
Saturn. Any moon imprudent enough to come closer to its parent
than a certain danger limit, known as the ' Roche limit ', is liable
to be torn to pieces by the tremendous, tide-raising gravitational
forces. So far Saturn is the only planet to have acquired a set
of rings in this way, but the inner moon of Jupiter is approaching close to the Roche limit, as also is Mars' Phobos, although
it is probably too small to be broken up. One day even our own
moon is expected to suffer the same fate, but this event will
not occur for many thousands of millions of years.

In addition to his rings Saturn has nine other moons, all of
which have been given names. In order outwards they are
Mimas, Enceladus, Tethys, Diona, Rhea, Titan, Hyperion, Iapetus
and Phoebe. Titan, the largest, is some 3,000 miles across and
can be seen with quite a small telescope. It takes about fifteen
days to make each circuit and has the distinction of being the
only moon which is known to have an atmosphere. Mimas, the
innermost moon, is only 30,000 miles from the edge of the rings
and it is believed to be largely the gravitational pull of this little
fellow, in conflict with that of the planet itself, which is responsible for keeping Cassini's division in being. Should a particle
stray into this gap it would very soon be swept out again, either
to one side or the other. Such are the intriguing ways in which
the forces of nature conspire.

URANUS

The sun, the moon and the five naked-eye planets were known to antiquity. In some languages they give their names to the days of the week. Although the telescope revealed the presence of a number of moons, no one ever suspected that there might be any more planets. Suddenly, on the night of 13th March 1781, William Herschel doubled the known size of the solar system. He discovered the planet Uranus, which travels round the sun at a distance of 1,782 million miles (nineteen Astronomical Units), taking eighty-four years for the journey.

Although, not surprisingly, many people regarded the discovery of Uranus as a stroke of fortune, Herschel himself would not agree to this. He had for some years been systematically examining each part of the sky, and, in his own words, ' . . . it was, that night, its turn to be discovered'. In fact Uranus can just be discerned with the naked eye; also he had been seen and recorded as a star on a number of occasions before 1781. Herschel, however, perhaps because of the superiority of his telescopes, realised at once that this was not a star. His first thought was that it might be a comet, but quite quickly he recognised it for what it really was.

Uranus is another giant planet, with all the forbidding characteristics of Jupiter and Saturn. He is, however, somewhat smaller than these two monsters, with a diameter of some 29,300 miles, and 'weighs' only the equivalent of 14·7 earths. As he rotates in about 10¾ hours, he has the typical oblate figure of a giant. In a telescope Uranus shows a faint greenish disc owing to an abundance of methane, and it seems likely that in structure, too, he is a typical giant with a solid core surrounded by a sheath of ice and a dense atmosphere of hydrogen with the same unpleasant additions.

Herschel thought he had discovered six moons of Uranus,[1] but only two of these have stood the test of time. Others, however, have been discovered since — the most recent, Miranda, in

[1] An orrery in the Old Museum, Glasgow, made during Herschel's time, shows Uranus with six moons.

1948 — and Uranus now boasts five which have been named from a variety of sources, Oberon and Titania, Ariel and Umbriel, and finally Miranda. Even Titania, the largest, is only about half the size of our moon.

Quite the most extraordinary aspect of Uranus is the inclination of his axis. He seems to have 'toppled over' on to his side and now proceeds with his north pole 8° *below* the plane of his orbit. The sun thus shines on each pole in turn for forty-two years at a time and his moons, which continue to lie in the plane of his equator, appear to move roughly at right angles to his orbital motion. Here, then, is yet another puzzle which has not yet been explained.

NEPTUNE

The discovery of the next planet, Neptune, is one of the most intriguing examples of mathematical detective work in the history of astronomy. From earlier and previously unrecognised observations of Uranus it was possible to determine his orbit with reasonable accuracy and for some years he appeared to behave as expected. Early in the nineteenth century, however, he began to deviate and in 1830, fifty years after his discovery, Airy, who was then Astronomer Royal, recorded 'Uranus has now deviated by the intolerable amount of almost two minutes'!

The most likely reason for the discrepancy was that Uranus was being perturbed by the pull of an unknown planet whose orbit lay even further away. In 1841 a young undergraduate at Cambridge, John Couch Adams, set about the intricate and formidable task of calculating its position. His homework took him four years, and when he had completed it he sent his predictions to Airy, telling him exactly where he expected the missing planet to be. Airy was sceptical of the idea and, incredible as it now appears, *nothing* was done to organise a search. In the meanwhile, in 1845, a Frenchman named Leverrier decided to tackle the problem independently, not knowing of Adams' work, which was by then lying idle at Greenwich.

Still worse was to come. In the following year Sir John Herschel (son of Sir William) wrote ' . . . the past year has given us the prospect of a new planet. We see it as Columbus saw America from the coast of Spain '. Excitement was in the wind. Leverrier completed his work and also passed his results to Airy who seems to have been delighted that they agreed with those of Adams, and eventually decided to take some action. He passed Adams' calculations to an astronomer at Cambridge where a large, new telescope had just been set up. Unfortunately this gentleman, because he did not work out the planet's up-to-date position, actually saw it twice but did not recognise it!

Leverrier, in the meanwhile, had also passed his results to a German named Galle, in Berlin, who received them on 23rd September 1846. He looked for and found the missing body the very same night because he just happened to have a map of that portion of the sky available for comparison. Although history has allowed Adams an equal share in the glory of the discovery, his could have been the full honour, had the Cambridge astronomers been a little more diligent.

The new planet, Neptune, has turned out to be very nearly a twin of Uranus, both in size and appearance, and is about one and a half times as far away. It was the first planet not to fit exactly into place according to Bode's Law. The average distance of Neptune is 2,792 million miles (or about 30 Astronomical Units) from the sun, and his journey takes him about 165 years. Only two moons appear to accompany this new planet, Triton and Nereid, the latter being the faintest satellite that has yet been discovered, a tiny, remote little world probably no more than 200 miles across.

PLUTO—THE ENIGMA

And so we come to Pluto, the furthest known outpost of the solar system. The discovery of Neptune had given most dramatic proof of the validity of Newton's theory of gravitation, in forecasting the position and size of a hitherto unknown planet. Towards the turn of the century, however, the new planet itself appeared to be misbehaving. Could there be yet another body, even further out, which was responsible for these perturbations?

Once again we can imagine the excitement beginning to mount. The necessary exploratory calculations were made independently by Percival Lowell and another eminent American, W. Pickering, but although a long and careful search was instituted, for many years nothing was found. This was a grave disappointment. Although Lowell himself had died in 1916, hope was never abandoned at his own observatory. In 1929 with the help of a new 13-inch refractor searching was resumed, and eventually, after months of arduous toil which involved examining hundreds of photographs, the missing planet was discovered in March of the following year, very close to its predicted position. It turned out to be something of a surprise.

Situated beyond the four giant planets, Pluto proved to be entirely different. A solid little body, in size the twin of Mars, he moves in the most eccentric orbit of all the planets. His mean distance puts him about forty times as far from the sun as is the earth, and although at times he ventures well *inside* the orbit of Neptune, his aphelion distance is nearly fifty Astronomical Units. However, his orbit is so steeply inclined to that of Neptune (about 15½°) that the two could never collide. (See Figure 9.) It has recently been suggested that this queer little misfit might once have been a satellite of Neptune, which, owing to perturbations by other giant planets, may somehow have escaped and gone into an orbit round the sun on its own. Pluto is certainly not unlike Triton in size and appearance and this hypothesis might well account for his apparently odd behaviour. Alternately he may be just one of a number of smaller planets whose combined effect could produce the observed perturbations of Neptune. Although further searches have been made, no further bodies have so far been discovered.

Very little is known for certain about Pluto. This planet has no perceptible atmosphere nor would we expect it to possess one necessarily, because of its small size and low temperature. Most gases, with a possible exception of hydrogen and helium, would be frozen out. His surface temperature is certainly very low, probably of the order of $-200°$ C. As we think of him pursuing his lonely path around the sun, moving at only 2·9 miles per second and taking approximately 248 years for the journey, we can thank our stars we were not born on Pluto in his perpetual winter!

THE LESSER BODIES OF THE SOLAR SYSTEM

Bode's Law had predicted, in 1772, the presence of a planet between Mars and Jupiter, but it was not until 1801 that anything was found which fitted into the gap. This was Ceres, 485 miles across and by far the largest of the minor planets. Its diameter is about one-fifth of that of the moon. In the following year another was discovered, Pallas, 304 miles across, and in 1804 Juno, only 118 miles across. Three years later came the discovery of Vesta, 243 miles in diameter, about the size of New York state. At a favourable opposition, this body can approach the earth close enough just to be visible to the naked eye. These new bodies became known as minor planets or *asteroids*, because of their star-like appearance. All four can be picked out with the aid of a small telescope.

After 1807 nothing further was found for thirty-eight years, and then between 1845 and 1850 six more appeared, all under 100 miles across. The following year saw the introduction of the camera to astronomy and with its help the numbers rocketed. By 1870 one hundred and nine minor planets were known, by the turn of the century about three hundred and fifty, and now the number is well over three thousand. If a camera is trained on some part of the zodiac and, with the shutter left open for a few hours, is guided to keep pace with the apparent movement of the stars, some minor planets are almost bound to be revealed by the short trails of light they leave as they move against the background of 'fixed' stars. In 1938 a search made in this way for further satellites of Jupiter revealed not only two new satellites but also thirty-one new asteroids.

Although each one of these little bodies is a planet, we cannot regard them as likely places for man to visit in the future. That they are 'tumbling' through space is apparent from the fluctuations in brightness that some of them show. The gravitational force of the smaller bodies is so low that centrifugal force might be able to eject anything or anyone landing on the surface. The escape velocity is so low that even the largest of them could not possibly retain an atmosphere; neither is it possible that they could hang on to a tiny moon for this, too, would be bound to escape through perturbations by the major planets. It seems probable that the asteroids are barren, lonely little worlds and are destined to remain so throughout eternity — unless, of course, some accident should befall them.

Owing to perturbations by the major planets the orbits of the asteroids themselves are undergoing constant changes and it is often difficult to keep track even of those that are known. Certainly they no longer lie neatly between Mars and Jupiter as predicted by Bode. Icarus, in fact, has an orbit which takes it to within 18 million miles of the sun, well inside the orbit of Mercury and in keeping with his namesake in mythology who ventured too near to the sun and melted the wax on his wings. A large number of minor planets can approach close to the earth, and in 1937 Hermes passed only 400,000 miles away from us. Nevertheless, despite the vast number of flying mountains there are about, it is encouraging to know that the chances of a collision are still probably less than once in a hundred thousand years.

As we move on down the scale, with asteroids getting smaller and smaller, there may well be no clearly defined line to differentiate between them and the next type of object, the meteors. Asteroids we know of only when they are large enough to be seen, and these number probably only a few thousand. Meteors we only know of when they actually collide with the earth and burn up in the atmosphere, probably between 70 and 100 miles above the earth's surface. The occasional larger ones we see as 'fire-balls', but the remainder we describe simply as 'shooting stars'. Each one is in reality a tiny planet that has been pursuing its independent orbit round the sun for millions of years. The meteors may have originally resulted from the disruption of two colliding asteroids. Most of the tiny micro-meteorites are so small that they could only be seen under a microscope, and it is probable that several million of these enter the earth's atmosphere every day.

As might be expected, meteors are more frequently seen after midnight than before as, owing to the earth's rotation, we are then on the 'forward' half of the earth and better placed to sweep them up in our passage. In recent years radio astronomers have also been able to 'observe' them by day, and have found them of great assistance in their study of the upper atmosphere of the earth.

Occasionally, as we saw in Chapter II, larger meteors do penetrate right through the earth's atmosphere before burning up, and in this way bring us our only first-hand information about the material of which other worlds are made. Study of the radioactive substances in some of these meteorites has enabled scientists to estimate the probable age of the solar system, in just the same way as the geologists have done for the age of the earth, and this they have put at about 4,600 million years.[1]

Whilst most shooting stars originate from random meteors, there are others which occur in definite showers at the same time each year, when the earth's orbit apparently crosses a clearly defined stream of particles. Some of these streams have been successfully associated with the paths of comets and it seems probable that the particles are the debris resulting from their passage. Meteor showers such as these appear to radiate from a particular part of the sky just as straight, parallel railway lines coming towards us from some distance away, appear to fan out, or radiate from a vanishing point. The meteors in a shower may be considered as moving in parallel tracks around the sun, so that as they approach us the trails they leave in the atmosphere also appear to fan out, and the vanishing point is known as the *radiant* of the shower. It is normally named after the constellation in that part of the sky. Typical showers which can be expected to give spectacular displays are the Perseids in August and the Geminids in December, both of which may produce as many as fifty or sixty visible trails in an hour. At one time the Leonids (mid-November) were also wonderfully active,

---

[1] Recent researches have identified traces of possible cell structures which might have been organic in origin. Although regarded by some as evidence of extra-terrestrial life, this cannot yet be regarded as proven.

and during the famous showers of 1833 and 1866 were described as 'falling like snow-flakes in a blizzard'. Alas, this shower is now much dispersed and is but a shadow of its former self!

COMETS

Of all heavenly bodies the comet is perhaps the most appealing to the imagination, although it is very seldom that a really spectacular one is visible to the naked eye. Nevertheless those that have been seen have undoubtedly created a tremendous impression, as through the ages it appears that comets have been regarded as omens of disaster, often foretelling the over-throw of kings and nations. Many events in history have been associated with their appearance, and the famous Bayeux tapestry portraying the Norman Conquest in 1066 features Halley's comet as a most terrifying object.

At one time comets were known as 'hairy stars',[1] but to-day of course we know them to be not stars, but members of the solar system whose movements are still closely controlled by the sun. They are rebellious members for they appear to obey none of the normal rules of the road. The planets all move in the same direction round the sun; the comets move equally in either direction. The orbits of the planets are almost circular and lie very nearly in the same plane; those of the comets are very elongated and may lie in any plane.

Although a comet is normally thought of as having a long, flowing tail, it is probable that for most of its life it will not have a tail at all. Its nucleus, or *coma*, is believed to consist of a rather loose, spongy conglomerate containing lumps of stony iron bound together by various ices and frozen gases. Methane, ammonia and carbon dioxide are most probably present. It is only when the comet approaches the sun and these gases are heated and driven out by the *solar wind* that it possesses a tail. For this reason the tail of a comet does not stream out behind it like the smoke from a railway engine but *always points away from the sun*. It may stretch for millions of miles through space, but is extremely rarefied. Stars have

[1] *Kometes*=long-haired (Greek).

appeared quite undimmed by a comet's tail and on occasions when the earth has passed through the tail of a comet there appears to be no record of its effect ever having been detected.

Many comets are periodic, and come back time and again at regular intervals which may vary from about three years up to a century or more. One of the best known, and one of the most spectacular, is Halley's Comet (see Plate 14), which has been seen many times through the ages, and whose appearances can be traced back as far as 466 B.C. It was Edmund Halley, Flamsteed's successor as Astronomer Royal, who first suggested that these unexplained intruders might be periodic. While studying the orbits of a number of selected comets he noticed the similarity between those of 1531, 1607 and 1682, and with great boldness forecast that the comet would return in 1758. Halley himself died in 1742, but the comet which now bears his name was found again on Christmas Day of 1758, a great triumph for English astronomy. It has a very eccentric orbit which, from a perihelion distance inside the orbit of Venus, takes it out beyond the path of Neptune. Its last appearance was in 1910 and with a slightly variable period of 75-76 years, depending on the perturbations of the major planets as it crosses their orbits, it is expected back in our skies again about 1986.

Where the comets come from we do not know. There is an interesting modern theory which suggests that these rather nebulous 'left-overs' may originate in the outer fringes of the solar system where it is thought there may be many millions of them. The ones we see are merely the occasional throw-outs that have been diverted from their orbits and venture in towards the sun. Sometimes their orbits are so disturbed by the other planets that they escape from the solar system altogether and are never seen again. Certain it is, however, that no comet can exist for long, when considered on the cosmic time scale, once it approaches nearer to the sun than the distance of Jupiter, for it starts losing its matter and must very quickly become dispersed. A life-time of a few thousand years is probably the most that it could expect.

HOW DID THE SOLAR SYSTEM COME INTO BEING?

It is a peculiar fact that we probably know less about the origin of our solar system than we do about the early development of the universe itself. Over the past two centuries, however, although there have been many widely differing ideas, they all appear to have fallen into one of two categories. One type suggests that the system was formed from some kind of nebulosity or cloud of dust and gas either at the same time as the sun or at a later period; the other that it was the result of some unique accident such as the close approach of another star or the break-up of a twin of the sun.

To be valid any theory must be able to explain the many regular features we have observed, the flatness of the system, the nearly circular orbits in which the planets move in the same direction, the moons, the differing types of planet, terrestrial and gaseous giants, and possibly the strange coincidence of Bode's Law.

Late in the eighteenth century Laplace suggested that the sun was formed from a vast gas cloud which, as it contracted, started to rotate and shed rings of gas which condensed to form the planets. These in turn, as they contracted, would repeat the process, giving birth to their moons. It was a simple hypothesis and it held pride of place for over 100 years. Eventually, however, it was shown to be untenable because the sun, in the process of contracting, would have been expected to start spinning at a tremendous rate, and this it clearly does not do. It is difficult to explain, on this hypothesis, how the other members of the solar system, while possessing only one-thousandth part of the total mass of the system, managed to possess 98 per cent of its angular momentum. Further, it appeared on closer investigation that the rings of gas would be more likely to disperse than to form into planets.

Next came the various encounter theories in which it was suggested that the close approach of or even collision with a passing star had been responsible for the eruption which had

[1] Page 35.

given birth to the planets. Sir James Jeans envisaged a cigar-shaped filament of gas being drawn out of the sun, thickest in the middle, where it formed Jupiter, and tapering towards the ends to form the smaller planets. This was an attractive theory in spite of the fact that, as we saw earlier,[1] the chances of such an encounter are probably no better than once in ten thousand million years. It had the advantage of making us possibly unique in the universe. Once again, however, it had to be abandoned as it was shown that any such resulting gas filament would be at such a fantastic temperature that it would rapidly blow itself to pieces and disperse, and could not possibly condense into planets.

Modern theories, postulated within the past two decades, have reverted once more to the nebulosity type. It now seems likely that our sun at some time acquired a vast exterior gas cloud, either as a direct result of its own formation or by sweeping it up during its passage round the galaxy. Such a cloud would have a tendency to condense into a number of separate globules, which have been called proto-planets, whose size would depend on the density of the cloud in that region. A German named Von Weizsacker has suggested that the condensations would individually start to rotate as the result of friction between her edges, something like a system of ball bearings. This might result in the transfer of angular momentum from the centre toward the edge. In this way he endeavoured to account for the present slow rotation of the sun. Much of the mass of the proto-planets would be lost in the process of formation, so that the planets would eventually settle down to their present sizes. He accounted for asteroids and comets by the suggestion that certain low-density regions could form only small bodies and not proto-planets.

Although this theory is not now generally accepted, most astronomers believe that some forces, perhaps of electromagnetic character, played an important role in the development of planetary systems. Gradually, astronomers have come to reject the catastrophic ideas of Sir James Jeans in favour of some natural process, which would make planetary systems widespread throughout the universe.

[1] See page 35.

This aspect is of particular significance when we come to consider the chances of life elsewhere in the universe. If our solar system resulted from a rare event, the chances of other systems and other life are small. If, on the other hand, the processes of its formation are likely to occur naturally in the order of events in the universe as a whole, who can say how many families of planets there may be among the myriads of stars in the heavens?

CHAPTER IV

—••€)(3••—

THE STARS ABOVE US

> . . . *as stars to thee appear*
> *Seen in the galaxy, that milky way*
> *Which nightly as a circling zone thou seest,*
> *Powder'd with stars.*
>
> MILTON : *Paradise Lost*

FEW VISIONS can excite our wonder more than a clear starry sky when the air is clean and the twinkling diamond orbs seem to pierce the darkness with their brilliant shafts of light. Its beauty is breathtaking. Standing under such a heaven we can forget the turmoil of civilisation and be at peace with the world.

In this chapter we are going to look at the stars mainly as individuals and shall be seeing what a vast range of characteristics they possess. But really to know them as friends requires more than just this. It includes the ability to recognise them from their position in the sky, to know their names, to fit them into their patterns; it means learning their stories and enjoying their presence. To do all this there is really no substitute for getting out under the stars and picking them out for ourselves, for in so doing we absorb at the same time the fascination and wonder of the universe around us.

Hipparchus mapped the sky in 120 B.C. into 48 constellations.[1] Other areas of the sky existed but they were too far to the south for him to be able to see them. To-day, by international agreement the sky is divided into eighty-eight areas, each named after the constellation it contains. The limits of these areas thus have their origins in the dawn of astronomy and are mixed up with the mythology and folk-lore of many

---

[1] *Constellatis* = studded with stars (Greek).

ancient civilisations. The patterns we see in our sky are virtually the same as those observed by the Chaldeans, the Pharaohs and the Phoenicians, for although the stars are moving at tremendous speeds through space, their distances are so enormous that such movement is not discernible with the naked eye during the span of a life-time, or even over a thousand years. If the astronomers of Ancient Greece could return to earth to-day they would certainly be bewildered by our daily lives, but they would still be completely at home amongst the stars. The figures of the gods and goddesses, the giants and heroes that they knew, would be as familiar to them now as they were 2,000 years ago. For convenience, a list of the constellations is given in Appendix II.

Many of the brighter stars have individual names by which they are popularly known (e.g. Sirius, Arcturus, Polaris, etc.), but it would clearly be impracticable to extend this to cover all the naked-eye stars, which on any clear night may number 3-4,000, let alone the myriads of fainter ones revealed by a telescope. The system employed by the early astronomers was to call the brightest star in each constellation α (alpha), the next brightest β (beta), and so on right through the Greek alphabet, following the letter by the genitive of the constellation name. Thus γ (gamma) Orionis is the third brightest star in the constellation of Orion. When the Greek alphabet is used up Latin letters are used, and as we move on down the scale other methods are introduced, often referring to numbers in a particular catalogue (e.g. H. D. = Henry Draper), but these need not concern us at this stage.

It has often been said that the pattern of the stars in a constellation bears little resemblance to the picture it is reputed to portray. Those who feel thus are missing much of the fascination of star-gazing. A dark sky is essential, since most of the constellations rely to some extent on the filigree of faint stars to complete the outline. Difficult as it often is nowadays to divorce ourselves from the glare of artificial lighting, it is worth striving for. The ancients had no such handicap. To emulate their imaginative picturing of the constellations, we also must endeavour to free ourselves from the brilliant lights of our cities. There are many little books[1] to help the beginner recognise the constellations. A star pattern, once picked out, becomes a friend for life.

The constellations whose names are most familiar will undoubtedly be the twelve which form the belt of the zodiac, that 'Circle of animals', within which lie the paths of the sun, moon and planets. There is a little jingle by Isaac Watts which readily recalls them to memory:—

> *The Ram, the Bull, the Heavenly Twins,*
> *And next the Crab the Lion shines,*
> *The Virgin and the Scales;*
> *The Scorpion, Archer and the Goat,*
> *The Man who pours the water out,*
> *The Fish with glittering scales.*

Unfortunately not all these constellations are particularly prominent, so that as the night sky gradually changes through the year the outstanding constellations which we associate with each season do not necessarily include these in the zodiac. As spring gives way to summer, so the Lion, the Virgin and the Herdsman give way to Hercules, the Swan, and the Scorpion. When summer in turn makes way for autumn, so we find the winged horse, Pegasus, dominating the sky in company with the fair Andromeda, the Goat and the Fishes. Finally when the long winter's nights are with us once more our sky is adorned by that glorious tableau of Orion, the Hunter, who, accompanied by his Hunting Dogs, wages his age-old battle with Taurus, the Bull. Spectators to this scene are the Heavenly Twins and Perseus, flanking the fine constellation of the Charioteer. These few constellations are sufficient with which to make a start. Later on, others will be found to fill in the gaps, and gradually the whole mosaic of the sky will fall into place.

The constellations named above are among the most important visible from the latitude of the United States. As we have already noticed, however, there are certain stars in the vicinity

---

[1] *A Field Guide to the Stars and Planets* by Donald H. Menzel, Houghton Mifflin Publishers; *The Friendly Stars* by Martin and Menzel, published by Dover.

of the North Pole star which do not do this, but are always above the horizon. We call them *circum polar stars*. The seven stars of the Big Dipper, which the British call the Plough, form part of the constellation of the Great Bear. In the early spring this group appears high overhead. In the autumn it lies low along the northern horizon, as if the Bear were about to hibernate for the winter. On the opposite side of the pole is a conspicuous ' W ' of stars belonging to the constellation of Cassiopeia. These two star groups circle daily around the star Polaris, like the hands of a giant celestial clock. Polaris, the North Star, lies at the end of the tail of the Little Bear. Cassiopeia is accompanied by her husband, Cepheus, the Emperor of Ethiopia, as they both watch their unfortunate daughter Andromeda (which is not a circum-polar constellation) chained to a rock as a sacrifice to the terrible sea monster (Cetus, which is still further to the south). Yet another conspicuous northern constellation is the Dragon whose coils can be traced round a full half circle of the northern sky.

The patterns and their stories would fill a book on their own.[1] Whenever the night is really dark and clear it is also possible to see that wonderful silvery band of light, the Milky Way, running like a great luminous river across the sky. We remember that the Milky Way also consists of stars, thousands of millions of them, but all too far away to be seen as individuals with the naked eye.

By taking our minds back to the model of our galaxy as a bun, ten miles across, we can see how this effect comes about. Our sun, we recall, is situated in the central plane about three miles from the middle of the bun, and all the naked-eye stars lie within about 500 yards of it, thus obviously belonging to our own galaxy. If we look from the earth towards the top or bottom of the bun we are able to see practically out to the edges, where the bun itself borders on the almost empty regions between the galaxies. The space between the individual stars thus looks empty. If, on the other hand, we look at right-angles

[1] *Patterns in the Sky* by J. D. W. Staal (published by Hodder and Stoughton) deals very pleasantly with this subject.

to this direction and in the plane of the bun, the space between the stars is filled with the diffuse light of the millions of stars which lie beyond the 500 yards but are still within the limits of our galaxy. Their combined effect produces the Milky Way as a glowing circle around the sky.

When we look towards the middle of the bun we would expect this effect to be even more marked than when we look towards the nearest edge and this, too, is very apparent. Those regions of the Milky Way which lie in the direction of the constellations of Scorpius and Sagittarius are far denser than those which run through Auriga and Gemini. They show themselves as tremendous star clouds in the sky (Plate 15), and astronomers are now convinced, for this and other reasons which we shall encounter presently, that the centre of our galaxy does indeed lie in the direction of the constellation of Sagittarius.

THE DISTANCE TO THE STARS

We shall begin our study of the stars themselves by looking at the question of distance, for in so doing we shall learn about many of their basic characteristics. So far we have seen how it has been possible to *measure* the distance of the moon and to calculate the distance to the sun by means of shorter measurements taken on asteroids and planets. We saw in our model that the distance to the nearest star, Alpha Centauri, was a quarter of a million times the distance to the sun, or four and a third light-years, but how do we measure such a distance? Radar is no longer of any assistance. Even if we could transmit a radio signal of sufficient strength to go to Alpha Centauri and return as an echo, we should still have to wait nearly nine years for it to make the round trip! A light-year, let us recall, is the distance travelled by a ray of light *or a radio wave* in a year, and amounts to almost six million million miles.

Once again it is necessary to revert to the surveyor's method of measuring out a base line and observing from either end of it. In this case the base line used is the diameter of the earth's orbit round the sun, approximately 186,000,000 miles in length. If we observe a ' nearby ' star in relation to the distant stellar background at intervals six months apart, a shift will be apparent owing to the earth's movement during the interval, and its amount will depend directly on the distance of the star observed. (Figure 13.)

This is the method of parallax and can be demonstrated by the following very simple experiment. Holding a finger out in front of you, close each eye alternately. The finger will appear to jump, and the closer it is held to the eyes the greater the jump

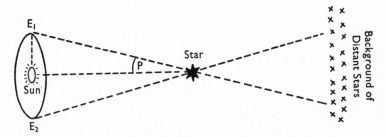

Fig. 13. Method of obtaining trigonometrical parallax of a nearby star. The star is observed when the earth is at E1 and E2. (A third observation, when the earth gets back to E1, is also necessary in order to eliminate the effect of the star's proper motion.)
The parallax of the star is the angle P, subtended by the Astronomical Unit.

will be. In theory it is a comparatively simple matter to apply this method to the stars, but because of the vast distances involved, the 'jumps' will be very small indeed. In our experiment the distance between our eyes (approximately $2\frac{1}{2}$ inches) represents the diameter of the earth's orbit. On this scale our finger (or somebody else's!) would have to be held five miles away in order to represent the nearest star! Clearly its apparent movement would be very small indeed.

Work on measuring stellar distances by the trigonometrical method, as it is called nowadays, is done entirely by photography, but it is interesting to note that the first successful measurements were made as long ago as 1838, before the invention of the camera. Bessel in Königsberg measured the distance of the star 61 Cygni (eleven light-years), while Her Majesty's Astronomer Royal at the Cape of Good Hope, Thomas Henderson, determined the distance of Alpha Centauri (4⅓ light-years), still the closest known star. In the same year another German, Struve, obtained a reasonably accurate distance for Vega which is twenty-seven light-years away, a remarkable achievement.

These figures themselves, when expressed in light-years, do not sound very impressive so, lest we should tend to think of the nearest stars as ' close ', let us digress in order to consider a simple example in terms of space travel. A space-rocket which could travel at 100,000 miles per hour (probably not an unreasonable speed for the future, when we remember that it starts with a 65,000 m.p.h. boost as a result of the earth's movement) would take two and a half hours to travel the distance to the moon, about six weeks to the sun, but *thirty thousand years* to the nearest star ! Space travel to the stars is quite clearly ' out ' for the time being.

When discussing the Astronomical Unit[1] we saw that astronomers tend to talk more frequently of the minute angles they actually measure than of the resulting distances. They thus refer to a star's *parallax* which, as can be seen from Figure 13, is the angle subtended at the star by the Astronomical Unit. When they do introduce distance it is more usually in terms of the *parsec*, which is the distance at which the Astronomical Unit subtends an angle of one second of arc, and is equal to approximately 3·26 light-years. The angles measured are thus extremely small and the parallax of the nearest star, Alpha Centauri, is only 0·76 seconds of arc. Nevertheless by modern photographic methods parallaxes of only 1/100th of a second have been measured, representing a distance of 100 parsecs or 326 light-years, and already about 10,000 stars have had their distances determined in this way.

We will not dwell too long on the difficulties encountered by the astronomer in this work for they are in keeping with the challenge he faces. Obviously he is measuring quantities he cannot see, the size of the image on his photographic plate is larger than the shift it experiences, the photographic emulsion

[1] See Page 92.

may move during developing, causing false displacements, there may be flexures of his telescope; all these items and many others must be allowed for. He does not claim, therefore, that his measurements are precise and would recognise that at a distance of 100 parsecs they may be as much as 50 per cent in error. Nevertheless this is not a bad start. If we revert once more to the experiment with our eyes, the finger is already 375 miles away! Pluto, the outpost of the solar system, is approximately five light-hours from the sun, but we have now penetrated to over 300 light-years. This, however, is the limit by this method until man can set up an observatory on one of the superior planets, where the length of his base line will be correspondingly increased. Another method of extending our base line by using the sun's own movement through space is discussed later in this chapter, but in the meanwhile if we wish to take our measurements a stage further we must examine some of the characteristics of the stars, quite the most obvious of which is their apparent difference in brightness.

STELLAR MAGNITUDES

From earliest times it has been the custom to grade the stars into *magnitudes*. Hipparchus used this method in the first star catalogue, and Ptolemy explained it in the *Almagest*. A 1st magnitude star was a very bright one (Vega was taken as the standard), and stars were then graded down to the 6th magnitude, which was a star that could just be seen with the naked eye. When precise measurements of the light from the stars became possible it was discovered that a 1st magnitude star gave out approximately 100 times as much light as one of the 6th magnitude. Dividing up the gap, we find that a 5th magnitude star is roughly two and a half times as bright as one of the 6th magnitude; a 4th magnitude star two and a half times as bright as a 5th, and so on. The precise ratio has been set at 2·512, which, when multiplied by itself five times, comes to exactly 100. Stars which fall between two magnitudes are given decimal values (e.g. Polaris is mag. 2·2). When the telescope reveals stars fainter than those which can be seen with the naked eye, these are catered for by extending the scale downwards, always by the same ratio. Already objects of the 24th magnitude have been photographed with the 200-inch Hale telescope and these are so faint that it would require $(2 \cdot 512)^{23}$, or 1,580,000,000 of them, to be as bright as one 1st magnitude star. If, by means of photo-electric methods, it becomes possible to increase its sensitivity 100-fold, this would enable objects of a further five magnitudes, i.e. down to magnitude $+29$, to be studied. The scale can also be extended in the opposite direction to include very bright stars of magnitude 'Zero', and beyond this to negative values, the ratio between one whole magnitude and the next always being $2 \cdot 512$. Sirius, the star which appears the brightest in the sky, has a magnitude of $-1 \cdot 52$, a full moon about $-12 \cdot 5$ and the Sun $-26 \cdot 72$ (equivalent to looking at approximately ten thousand million 1st magnitude stars all at once).

How can all this help us in our quest for distance? Quite obviously stars can differ in brightness for two reasons; either their intrinsic brightnesses may differ, or they may be at greatly differing distances from us. The way in which these factors are used can be simply illustrated as follows.

Let us consider two electric lamps, both of the same brightness, say 50 watts, one placed at a distance of 100 yards from us and the other at 200 yards. Provided we know that their true brightness is the same we can quite simply compare their distances by observing their apparent brightnesses. If we now replace the further lamp by one of 200 watts (i.e. four times as bright) both lamps *will look the same*, because brightness falls off according to the *square* of the distance.[1] Thus, even if we had not known that the further lamp was one of 200 watts, we could have deduced this fact, provided we knew its distance. Alternatively, if we could only find a way of telling the *true* brightness of a lamp we could always compare this with its *apparent* brightness and so obtain its distance, no matter how

---

[1] Doubling the distance reduces the light received by a factor of four. Trebling the distance reduces the light received by a factor of nine. Quadrupling the distance reduces the light received by a factor of sixteen, etc., etc. From Pluto, at a distance of roughly forty Astronomical Units, the sun will appear only $1/1600$ ($= 1/40^2$) as bright as it does from the earth.

far away it was placed. These are the principles we are now going to use.

Provided a star is sufficiently close to have its distance measured by trigonometrical methods it is possible, from a study of its apparent magnitude, to work out how bright it really is. This value is called its *absolute magnitude* and, by definition, is the magnitude it would appear to have if it were placed at a standard distance of 10 parsecs (32·6 light-years). In other words, if every star were placed at a distance of 10 parsecs, they would all appear to us at a brightness equivalent to their absolute magnitude. Our sun, which has an absolute magnitude of only +4·6, would appear as a faint star only just visible to the naked eye. On the other hand the bright star Rigel, in the constellation of Orion, which is at a distance of about 650 light-years (roughly 200 parsecs) would, if brought in to 10 parsecs, appear as a brilliant object 1,500 times as bright as Polaris normally looks to us. Rigel has an absolute magnitude of −5·8, and is intrinsically more than 20,000 times brighter than our sun.

Thus if only we had some way of telling the absolute magnitude of a star just by looking at it, we could quite easily calculate its distance. But how can this be done?

SPECTRAL TYPES

We have already seen how it may be possible to tell the temperature of a star from a study of its spectrum. Let us now look a little more closely at this problem. A study of the spectra of stars has enabled astronomers to divide them into a number of separate classes, which have been given the letters O, B, A, F, G, K and M. Generally speaking O-type stars are the hottest and M-type are the coolest. Within each category further subdivision is possible by assigning numbers, from nought to nine. Thus our sun, which is designated as a ' Go ' (G zero) star, is relatively cool. Other categories also exist for stars which are in some way peculiar; W for Wolf-Rayet stars, the hottest type known, and R, N and S, for cool stars similar to type M, but with special characteristics such as strong bands of molecular carbon in types R and N and bands of zirconium oxide in stars of S, replacing the familiar bands of titanium oxide in the more abundant stars of class M.

About 1920 Hertzsprung and Russell plotted a diagram of the absolute magnitude of stars against their spectral type, for all the stars whose distance had been determined (i.e. those within about 300 light-years). Their result, which is illustrated diagrammatically in Figure 14, showed that by far the greater number of stars fell along a fairly narrow band of the graph which has been called the *Main Sequence*.

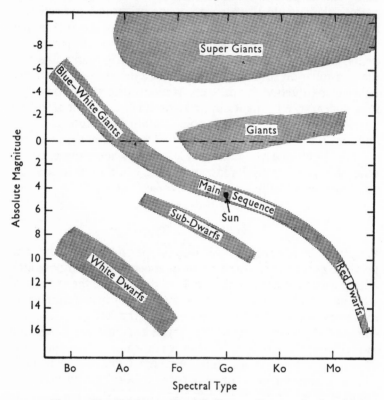

Fig. 14. The Hertzsprung-Russell diagram for stars in the vicinity of the sun, showing the principal groups into which the stars appear to fall. Spectral type is shown along the bottom of the diagram and absolute magnitude up the left-hand edge.
The sun appears in the Main Sequence as a Go star of absolute magnitude +4·6.

Here is the answer to our problem. If we assume that the laws of nature remain the same no matter how far into space we probe, it should be possible, by ascertaining the spectral type of any star, to fit it into its position in the diagram and so read off its absolute magnitude. As with the lamps, a simple comparison between this and its apparent magnitude should thus enable its distance to be determined.

This was a major break-through for here, at last, was a way of telling the true brightness of a star. At first it seemed as if certain ambiguities might still arise where a star of a particular spectral type could be fitted into the diagram in more than one place. For example a star of type Ko might be either a main sequence star or a giant; a star of type Bo might either belong to the main sequence or be a white dwarf. It is one of the triumphs of modern astrophysics that even this difficulty has now been resolved, and it is possible, with confidence, to distinguish between the spectra of each different class of star.

By means of the Hertzsprung-Russell diagram astronomers are, even now, extending their measurements deeper and deeper into space. The original three hundred light-years, which was the maximum possible by the trigonometrical method, has already been extended to several million. One of the main difficulties initially was to calibrate the 'ruler' over its full limits, for there are certain types of star, notably the bright O and B giants, which can be seen at great distances but which do not occur in the tiny volume of space open to our measurement by trigonometrical methods. To use them in the Hertzsprung-Russell diagram it was necessary first to determine the distance of at least one example. This problem has now been overcome by the use of yet another of the stellar characteristics we can observe, namely their motion through space.

PROPER MOTION

We have already referred to the fact that the stars are in motion, but that due to their immense distances no change in their position is apparent to the naked eye. A star's movement at right-angles to the line of sight is called its *proper motion* and the fact that stars have such motions was first noticed by Halley in 1718. He compared the position he had obtained for the star Arcturus with that given by Ptolemy in the *Almagest* almost sixteen centuries before, and found a difference of about 1°.

The proper motion of a star is very small and seldom amounts to more than a few seconds of arc a year. However, as its effect is cumulative, its determination becomes more accurate the longer the time between the observations. Like so many things in life, the best results are worth waiting for. It will be obvious that the closer the star is to us, the greater will be the effect of its speed on its apparent position. A motor-car passing our front gate is gone in a flash whereas an aeroplane flying high overhead, even though it is travelling faster, takes an appreciable time to cross our field of view. The closer stars are thus likely to have the largest proper motions and in order to find out how fast a star is moving at right-angles to our line of sight, we must know its distance.

Nowadays many thousands of stars have had their proper motions determined by photographic methods and the greatest value obtained is that for Barnard's star, a faint red dwarf only six light-years away, which changes its position by 10¼" per year. At the same time, of course, the observed star may also be moving rapidly towards or away from us, a movement which will not show up, as it will not affect the star's apparent position in the sky at all. It can, however, be found quite simply with the aid of the spectroscope. We have all experienced the well-known 'Doppler' effect which causes a train whistle to sound higher in pitch when the train is rushing at us and lower when it is going away. This occurs because, in the first instance, the movement of the train compresses the sound waves, making their wave-length shorter and their note higher; similarly it increases their wave-length as the train moves away and the note gets lower. Precisely the same thing occurs with the light from the stars, and the effect can be observed with the spectroscope by noting the way in which the Fraunhofer lines are displaced towards the blue end or the red end of the spectrum. A displacement towards the blue end tells us that the star is approaching us, because blue light is of a shorter wave-length than red; a displacement towards the red end tells us that the star is receding, and in either case the amount of the displacement tells us how fast the star is moving along the line of sight. This is called its *radial motion*.

To determine a star's true motion in space it is necessary to combine both the movements we have observed, its radial motion and its proper motion, and this can be done quite simply, as shown in Figure 15.

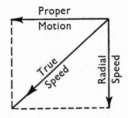

Fig. 15 Diagram to show how transverse speed, which is obtained from a star's proper motion and its distance, can be combined with radial speed to give the star's true motion in space.

DOES THE SUN HAVE A TRUE MOTION?

As the sun is a typical star it is natural to enquire whether it, also, is moving through space. When the true motions of other stars are plotted in a diagram it is found that *on the average* those which lie in one direction appear to be streaming towards us whilst those in the opposite direction seem to be streaming away from us. As it is most unlikely that this is really the case, it is fairly safe to assume that it is really the sun and its family which are moving, and in the direction of those stars which appear to be streaming towards us.

This direction is called the *apex* of the sun's movement, and the opposite direction is called the *antapex*. It appears that the solar system is racing towards a point in the constellation of Hercules, quite close to the star Vega, at a speed of about twelve miles per second, or nearly 400 million miles a year. To travel the distance to Vega would, however, take about 400,000 years, and by that time Vega itself will have moved many light-years out of our way.

DISTANCES FROM PROPER MOTIONS

If we know how fast a motor-car is travelling along a road running at right-angles to our line of vision, we can work out how far away it is by noting how rapidly it crosses our field of view. This is simply the reverse of the method of obtaining speeds from proper motions which we have already discussed. We have noted that *on the average* the stars in the direction of the apex appear to be coming *towards* us at twelve miles a second and those in the direction of the antapex appear to be going *away* from us at speeds of twelve miles a second, both being due to the sun's movement. In a belt half-way between these two directions, the stars on the average should appear to move straight *past* us at twelve miles a second. If we now assume that the average proper motion we do obtain represents this expected passing speed of twelve miles a second, then, as in the case of the motor-car, we can work out the actual distance of the stars concerned.

The essence of this ingenious method lies in the words ' on the average '. A large number of stars must be observed, in order that the effect of their own proper motions can be expected to cancel out, leaving only the residual apparent twelve miles a second caused by the sun's motion. This approach has proved of particular value in obtaining the distance of certain types of star of which we do not find specimens close enough to measure by the trigonometrical method. Once the distance of such a star has been found in this way its absolute magnitude can be determined. It can then be fitted into the Hertzsprung-Russell diagram and so used for determining the distances of similar stars which are even more remote.

Many other highly ingenious ways of extending our measuring rods have been devised, some of which will be mentioned later in this chapter. Before proceeding further, however, let us pause and look a little more closely at some of the other characteristics shown by the stars themselves.

It is well known that the colour of a light-source depends very largely on its temperature. As we heat a piece of metal it becomes red hot then rises through orange to yellow and finally to white. So it is with the stars. Their colour can be determined very accurately by photographic methods and this at once gives a very good indication of their probable surface temperature. There is thus a very close relationship between the colour of a star and its spectral type.

The coolest stars we know of are those of spectral types K and M, and the cooler the star, the redder is the light that it radiates. Three prominent examples stand out amongst our nearer neighbours: Antares,[1] in the heart of the Scorpion, seen low in the south during the summer, Aldebaran, the fiery eye of Taurus the Bull, and Betelgeuse pronounced (' Betelgerz '), a variable star in the shoulder of Orion, both of which look down from our winter sky. Aldebaran, which is really an orange star and slightly warmer than the other two, is known as a 'giant'. Antares and Betelgeuse are called 'super-giants'.

To understand what these terms mean we shall take, as our standard, our own sun. A typical giant star such as Aldebaran has a diameter about forty times that of the sun, and would fill the solar system about half-way out to the orbit of Mercury. Seen at the distance of the earth it would cover 20° in the sky, the size of a football held only two feet from the eye. We on earth could not, of course, survive with such a monster in our midst, but were we to have one of the super-giants in place of the sun our fate would be even more extraordinary. These stars are so enormous that we should find ourselves right inside them. Both Antares and Betelgeuse have diameters roughly three hundred times that of the sun and in volume they could contain about thirty million suns. Betelgeuse is also a pulsating star,

[1] The rival to Ares (Mars), Page 102.

which means that from time to time he puffs himself out so far that he would actually envelop Mars as well. His cycle of expansion and contraction is somewhat irregular. The reason for the variation is not fully understood, but the change of brightness amounts to a full magnitude. Both Antares and Betelgeuse are M-type stars with surface temperatures of about 3,000° C., only half the temperature we found on the photosphere of the sun. They are nevertheless brilliant because their tremendous surface area, all of which is radiating light, is about 100,000 times that of the sun.

Staggering as these stars may appear, they are completely dwarfed by the record holder, a star known as Epsilon Aurigae B which, if it were to replace the sun, would swallow the whole of the solar system as far out as the planet Saturn. It occupies the space that would be filled by 200 super-giants the size of Betelgeuse or 6,000 million suns. The outer regions of such a star are probably many thousands of times more rarefied than the air we breathe, and it seems possible that it is still in the process of forming.

These are the cool stars of the universe. If we turn now to the other end of the scale, the very hot stars of spectral types O and B, we find that the colour rule again holds good, for these stars are white or blue-white. Selecting another star in the constellation of Orion, which abounds in brilliant stars, let us look at Rigel, in the right leg of the great hunter. Rigel, too, is classified as a super-giant or sometimes as a 'Blue-giant', but he is of an entirely different type from Betelgeuse or Antares and really belongs at the top left-hand end of the Main Sequence (Figure 14). His brilliance is due less to his diameter, which lies probably between twenty and thirty times that of the sun, than to his fantastic surface temperature in excess of 20,000° C. It would take about 23,000 suns like ours to appear as bright as Rigel. Clearly this star must be using up its substance at a fantastic rate to produce this output of energy, and stars such as Rigel may be said to enjoy a short life and a gay one. Short, that is, on the cosmic time scale. His brilliance is likely to last no more than ten million years which, when we recall that our sun has already been shining for several thousand million years, makes us realise that stars such as Rigel must be, cosmically speaking, very young indeed.

Both Betelgeuse and Rigel are at distances from us of about 650 light-years, too far off, in fact, to have their parallaxes measured directly. A star which lies very much closer to us, and which falls between the two extremes of temperature we have considered, is Sirius the brilliant 'Dog Star'.[1] Sirius is less than nine light-years from us and is the closest naked-eye star visible from the United States. He is only a little larger than the sun, but being of spectral type A he is nearly twice as hot, with a surface temperature of about 11,000°C. His colour, as we should expect, is white and his 'luminosity' is equal to about twenty-six suns. Sirius has a very interesting little companion, known as Sirius B (or 'the pup') and the two form what is known as a *binary system*. These two stars are revolving round their common centre of gravity, as do the earth and the moon, but whereas Sirius A is brilliant, the pup can only be picked out with a very powerful telescope. He is, in fact, about three hundreds times fainter than our sun. This star is called a *white dwarf* and is interesting because here we have a star whose output of energy is no longer sufficient to balance his own gravitational force,[2] and his outer layers have thus collapsed inwards towards his centre. Although in size the pup is now little bigger than the earth, his mass is believed to be roughly equivalent to that of the sun. We can thus imagine how very densely packed must be the matter within him. A tumbler-full would weigh about fifty tons.

An unexplained 'wobbling' observed in the motion of Sirius itself led astronomers to suspect the existence of a faint companion revolving around Sirius, almost twenty years before he was actually discovered. Sirius B is extremely difficult to observe because of its closeness to the brilliant Sirius A.

Hitherto we have generally referred to the star which is closest to the sun as Alpha Centauri, but this now requires a little

[1] Sirius, in mythology, was Orion's favourite hunting dog. The star Sirius is the brightest one in the constellation of Canis Major, the 'Greater Dog'.
[2] See Page 159.

amplification. Alpha Centauri is, in fact, a multiple star system consisting of two stars of the same type as our sun, accompanied by a third star which is known as Proxima Centauri and is the component actually lying nearest to us. It is this little companion which we are to consider, for it is known as a *red dwarf*. These stars occupy the bottom right-hand corner of the Main Sequence. They are generally several times smaller than our sun but of much greater density. They are, however, by no means as dense as a white dwarf. Because of their red, cool surfaces, these stars are all faint and inconspicuous. None of them can be seen with the naked eye. Some have less than a thousandth of the luminosity of the sun. The fact that not many red dwarfs are known at present is due entirely to the limited volume of space in which we can search for them. In all probability they are among the most abundant stars in the sky.

It thus appears that a cool red star of type M may be either an inconspicuous dwarf or one of the largest giants in the universe, and the few examples we have looked at will suffice to illustrate the tremendous variety that exists among the stars in the sky. Whilst there are stars many thousands of times brighter than our sun there are also others which are thousands of times fainter. It may well be that the latter may eventually prove to be in the majority.

BINARIES

Many stars which appear to us as single points of light are in reality systems of stars with two, three, or possibly more components, and once we study the stars with instruments this turns out to be almost as much the rule as the exception. While there are a very few twin systems which can just be separated by an acute naked eye, the vast majority require some form of optical aid, and anyone with a small telescope can have endless hours of enjoyment picking them out. Frequently the two components appear in greatly contrasting colours as, for example, the stars of Beta Cygni in the eye of the Swan, which appear blue and yellow, or those in Beta Scorpii (the top claw of the Scorpion) which look mauve and green. An interesting star in the constellation of Lyra, close to the brilliant star Vega, is the faint 4th magnitude star, Epsilon Lyrae, which can be seen as a wide ' double ' with the aid of a pair of field-glasses. A small telescope, however, reveals that each component is itself a twin, and this star is the famous ' double double '.

Not all visual doubles are true physical systems. For example, Beta Cygni consists of two stars lying in the same direction but in no sense revolving around one another. The great importance of true binary systems is that they enable us to *weigh the stars*. The two components are revolving about their common centre of gravity under the influence of their combined gravitational pull. We might liken them to dancers spinning round and round each other, holding hands at arm's length. In accordance with Newton's Laws their distance apart and their period of revolution will depend directly on their combined masses. By careful observation over a period of time, these quantities can be determined and if the results are compared with those we know from observations within our solar system, they will enable the *combined* masses of the component stars to be found in terms of the mass of the sun. Finding out how much of this mass belongs to each component is difficult, as it is necessary to find out how much each star deviates from the mean movement of the system. As binary stars may take many decades or even centuries to revolve round each other this may take a long time. The period of Sirius A and the pup, for example, is about fifty years. Nevertheless the masses of several hundred stars have already been found in this way and it is interesting to note that although their sizes and brightnesses vary by a factor of many millions, with few exceptions their masses only range from a little less than our sun to perhaps ten or twenty times as much. In general we can say that most stars, in spite of their many other differences, weigh very much the same.

SPECTROSCOPIC BINARIES

Some binary systems are so compact that they cannot be separated, even with a powerful telescope. How then do we know that they are binaries?

The answer is that the spectroscope tells us so. We have mentioned previously[1] that the spectroscope can tell us whether a star is coming towards us or receding, and at what speed. When the light of a close binary star is studied with a spectroscope, it may happen that one component is approaching while the other is receding. The Fraunhofer lines from each will be displaced in opposite directions and will appear in the spectrum of the star as double lines, thus giving away the star's secret. The further the lines are separated the faster is the relative movement of the two components in our line of sight.

As an example of a binary star system let us take a look at an extremely interesting star, Castor (Alpha Geminorum), one of the Heavenly Twins. A small telescope shows Castor to be a splendid double star, which has a period of approximately 340 years. The spectroscope reveals that each component is a spectroscopic binary with a period of a few days. All four stars are comparable in brilliance with Sirius. As a final surprise there is also a faint red star, itself a spectroscopic binary consisting of two red dwarfs, which moves around the whole system with a very long period, possible of the order of a million years. Castor, which appears to the naked eye as a 'normal star', thus turns out to be a complex sextuplet which contains a little bit of everything!

ECLIPSING VARIABLE STARS ('WINKING' STARS)

When the plane of the orbit of a spectroscopic binary system lies near that of our line-of-sight, one star can pass in front of the other and it then appears as an *eclipsing binary*. One of the best known is *Algol*, in the constellation of Perseus, which was seen by the early Greeks as the wicked eye of Medusa, one of the Gorgons.[2] The two components of this star are both several times larger than our sun but they differ in brightness very considerably. Their period of revolution round each other is slightly under three days (sixty-nine hours). Whenever the fainter (and larger) star partially eclipses the brighter one, Algol

---

[1] See Pages 136–7.
[2] The name Algol comes from the Arabic ' Al Ghoul ', meaning ' The Demon ', a shorthand reference to the head of Medusa.

appears to do a slow wink, falling in brightness by a full magnitude and a quarter, and then recovering. The wink, which takes about ten hours to complete, can be clearly seen with the naked eye and the times of ' minimum ' are published each year in *Whitaker's Almanack*. Half a revolution after the main eclipse the brighter star will partially eclipse the fainter, causing a very minor wink, but this is normally only detectable with the help of sensitive devices for measuring the brightness of the sky, such as a photoelectric photometer.

Intriguing as these stars are to watch, their real importance lies, once more, in the amount that astronomers can learn from them. By observing the behaviour of these close binaries they can discover much concerning the sizes, masses, temperatures and brightness of the individual component stars, information which can often be applied later to other single stars with similar characteristics, for which such information might otherwise be very difficult to obtain.

OTHER TYPES OF VARIABLE STAR

We normally think of a star as shining quite steadily, and with the majority of stars this is undoubtedly true. Yet there are at least twenty thousand of them which are known to vary, not because of any eclipse cycle but because of true fluctuations within themselves. In the main they fall into three classes, long period variables, irregular variables and Cepheids, which are perhaps the most interesting and important of all, so we will consider these first.

The name ' Cepheid variable' comes from the star Delta Cephei which was the prototype of this class, more than a thousand of which are now known. Their basic characteristics are so distinctive that they can be recognised at once, rather as a lighthouse or an aircraft homing beacon can be identified, even at a distance, by the coded signal which it transmits. With the Cepheids the fluctuations in the transmitted light follow a well-defined pattern, rising abruptly to a peak and gradually falling away again at completely regular intervals (see Figure 16). The effect is rather like the glowing end of a cigarette at which steady and measured draws are being taken.

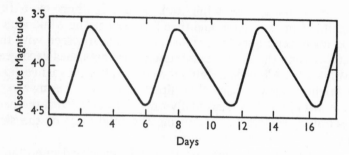

Fig. 16. The light-curve of a typical Cepheid variable. Note the rapid rise in brilliance followed, each time, by a more gradual decline.

Half a century ago an American astronomer named Miss Leavitt made a discovery in connection with Cepheid variables which was to open up a completely new avenue of approach to the task of charting the universe. Whilst studying the Cepheids in the small Magellanic Cloud (see Plate 16) she noticed that the longer their period of fluctuation, the brighter was their apparent maximum brilliance. As all the stars she was examining were within the same cloud, which was itself a great way off, they could for all intents and purposes be treated as all being at the same distance. It was thus possible to arrange them in order of their absolute magnitudes and it transpired that the brighter the star the longer did it take to go through its cycle of fluctuations. In terms of our cigarette, the longer the interval between the puffs, the deeper would our smoker inhale. Here, it seemed, was a wonderful yardstick for measuring distance. The business of timing the period of a variable star was simplicity itself, yet that apparently was all that was necessary in order to tell its maximum brightness. As we have already seen, once we know the absolute magnitude of a star, and can observe its apparent magnitude, its distance can be readily determined.

Only one thing remained and that was to *calibrate* the ruler. The actual distance of at least one Cepheid had to be measured, yet none was close enough for this to be done by trigonometrical methods. The problem occupied and frustrated astronomers for almost half a century. Various estimates were made, using a variety of methods, but each in turn had to be discarded in the light of later evidence. It was only in 1952, forty years after Miss Leavitt's first historic discovery, that Baade of the Mt. Wilson Observatory finally succeeded in establishing the true absolute magnitudes of these stars with a reasonable degree of certainty.

The great advantage of the Cepheid variables is that they are all giant stars visible over enormous distances. A Cepheid with a period of approximately three days is equivalent in brilliance to about 1,000 suns. Some are ten times as bright again. They thus provide astronomers with a wonderful series of ready-made mile-posts whose distances can always be read off, and with their help many of the remote regions of our galaxy have been mapped. Nor is this all. Cepheid variables have also been identified in certain of the nearer galaxies and so have provided a self-evident method of determining with considerable accuracy the distance of even these remote stellar oases.

The periods of Cepheid variables range from about one day to almost two months. There is, however, another class of variable star which is remarkably similar and which might almost be described as 'Cepheids in miniature'. These are the RR Lyrae stars. They appear to pulsate according to the same rules but with much shorter periods which range from under two hours to about a day. Their maximum brilliance is there-fore considerably below that of the Cepheids. The importance of these stars lies in the fact that they occur with great fre-quency in the globular clusters and so, as we shall see later,[1] are of great assistance in determining the extent of our galaxy. They are occasionally referred to as 'cluster variables'.

Most variable stars do not, however, operate with the clock-work precision of the Cepheids or RR Lyraes. Those classed as *long period variables* have fluctuations which may range from a few months to a few years, but they do not show the same period/luminosity relationship nor can their regularity be relied upon. One of the best known of these is the interesting star Mira Ceti, seen as the beating heart of the great sea monster,

[1] See Page 154.

Cetus.[1] Mira, meaning 'the wonderful' was discovered as a variable star by David Fabricius, in 1596, even before the invention of the telescope, and has a period of approximately 333 days. From being a star of the 9th or 10th magnitude, well below naked-eye visibility, it will gradually become more luminous and in a matter of some weeks rise through six or seven magnitudes to become a comparatively bright star in the sky. Thereafter it will very slowly fade, taking almost six months before it drops once more below the threshold of vision.

Several hundred long period variables are known and all are intrinsically very bright stars. Their study provides one of the fields of astronomy in which the amateur can play an important role, for it is only by continuous observation that the behaviour of these stars can be determined, and professional astronomers frequently cannot spare the time to take such observations. Although it is fairly certain that they are undergoing some form of pulsation, much still remains to be learnt regarding the internal mechanism of this strange class.

Some stars vary with a truly erratic period and are known as *irregular variables*. These, too, are mainly giant stars, of which Betelgeuse is a typical example,[2] but at the other end of the scale a comparatively recent discovery has shown that some of the very faint red dwarf stars should also be included under this heading. They have been named *red dwarf flare stars* and, while little is known of them for certain, it seems likely that their sudden bursts of activity are caused by something analogous to the solar flares we observe in our sun. Because of the normally feeble appearance of these stars, such outbursts are sufficient to increase their brightness to a very marked extent, and it seems that these elusive little stars may well have many more surprises in store for us.

NOVAE

From time to time a perfectly ordinary star in the sky may suddenly flare up and shine with many thousands of times its

[1] See Page 127.
[2] See Pages 139-140.

original brilliance. Such a star is termed a ' Nova ', meaning, literally, a ' new ' star. We now know, however, that this is really not a new star at all but an old one that has blown off its outer layers in a most spectacular fashion. Astronomers currently believe that the catastrophe has resulted from the star's having exhausted its normal source of energy deep in the core. As a result, such a star begins to ' burn ' hydrogen in the envelope. And, as this ' shell source ' of energy approaches the surface, the weight of the overlying layers can no longer control the atomic reactions. A tremendous explosion, analogous to the hydrogen bomb, results. Such stars are not particularly uncommon; an average year may produce a crop of perhaps a dozen in our own Milky Way, but most are so distant as to escape notice.

In just a few hours the star may brighten as much as 100,000 times, rivaling the brightest super-giants in the sky. But, after a few months or even weeks, the star fades and eventually returns to much of its original appearance, although the expanding envelope of ejected matter can often be detected for many years. These are the ' recurrent novae '. This group may in some way be akin to the regular variable stars.

SUPERNOVAE

We come now to one of the most spectacular of all the celestial fireworks, the Supernova. Whilst Novae are not particularly uncommon, Supernovae are very rare indeed. Only three are known to have occurred within our galaxy in the past two thousand years, the one of 1575, observed by Tycho Brahe, that of 1604, noted by Kepler, and one in 1054, of which we shall have more to say shortly. Whilst a Nova may achieve the brilliance of a hundred thousand suns, a Supernova may exceed the light of a hundred *million* suns; whilst a Nova quickly returns to its original state having lost only a minute fraction of its matter, in a Supernova we are clearly seeing *the death of a star* as it explodes and blows itself to destruction.

On 4th July 1054 the Chinese, who were always very careful chroniclers of everything that went on in the sky, recorded the appearance of a new and very bright star in the constellation of Taurus. This star suddenly blazed up and quickly became the most brilliant object in the sky. We are told that it outshone Venus and it may well have been visible even in the day-time. For several weeks the star remained a blaze of light, and then very slowly it began to fade away. Finally it was lost and search as they might, they never saw it again. To-day, more than nine centuries later, we can, with the aid of a good telescope, examine the remains of that great celestial catastrophe that appeared to those Chinese astronomers all that time ago. We call the result the Crab Nebula. (Plate 17.)

This interesting mass of stellar debris, which is entirely gaseous, is observed to have a very faint central star, probably a white dwarf, which is at a fantastically high temperature, and is probably the remains of the original star. The rest of it forms the lovely surrounding nebulosity which is still expanding at an enormous rate, and is gradually spreading its substance throughout a vast region of the galaxy. The light by which we observe the Crab Nebula is largely the result of the collisions between the gaseous filaments, although the very high frequency radiation from the hot central star is probably also causing this gas to fluoresce. In addition this remarkable object has proved to be one of the most powerful radio sources in the sky. As its distance has been found to be about 5,000 light-years, we will realise that the explosion observed by the Chinese in 1054 actually took place about 4,000 B.C.

During the past hundred years more than fifty Supernovae have been observed in other galaxies, and calculation has shown that on the average they might be expected to occur in any single galaxy about once in every three centuries. Their maximum absolute magnitude seems to be almost consistent at about −16,[1] so that they sometimes appear brighter than the whole galaxy in which they occur. Because of their great brilliance Supernovae can on occasions be used to determine the distances of galaxies which are too remote to be resolved into individual

[1] Supernovae actually occur in two types with maximum absolute magnitudes of −16 and −14. They can, however, be distinguished by the quality of their spectra.

stars, and for which Cepheid variables would thus provide no assistance. They give us yet another method of extending our measuring rods still further into space.

GALACTIC CLUSTERS

Leaving our study of the stars as individuals let us look now at the colonies or associations in which they are found. Two distinct and widely different types of star cluster are evident and are known as *galactic* (or 'moving') clusters and *globular* clusters. The former, as their name implies, are found only near the central plane of the galaxy; the latter are always far removed from it.

About 500 galactic clusters have been identified within our galaxy, varying in size from a few score members to several hundred, or even a thousand. Some of the nearest are naked-eye objects of great beauty, the Pleiades (Plate 18) being perhaps the best known of all. The Hyades, Praesepe (the beehive) in Cancer, the Double Cluster in Perseus and the lovely Hair of Queen Berenice are other examples which can be enjoyed without optical aid, although their fascination cannot be truly appreciated without a telescope.

The Pleiades (or the Seven Sisters) is a delightful little cluster in the constellation of Taurus, with its seven brightest members named Maia, Taygeta, Electra, Alcyone, Celoeno, Sterope and Merope. The story goes that they were the seven daughters of Atlas but that one of them, Merope, married a mere mortal and for this reason she now shines less brightly in the sky than her sisters!

A good eye may pick out as many as a dozen stars in the Pleiades, a pair of binoculars shows some fifty to sixty while a good telescope tells us that the cluster probably contains almost ten times this number. It is a scintillating group of stars that is truly beautiful to behold. Tennyson wrote of it:

Many a night I saw the Pleiads rising through the mellow shade,
Glisten like a swarm of fireflies, tangled in a silver braid.

A glance at Plate 18 surely shows us what Tennyson meant. The

stars are there as the fireflies, and the silver braid is the inter-
stellar dust we see so beautifully illuminated by the star-light
which it reflects.

Galactic clusters appear to be true families of stars that have
been born from the same 'cloud' of gas and dust. Their members share a common movement through space rather like an
enormous flock of birds and they are all at approximately the
same distance from us; but like most large families they have
all sorts amongst their number. Many of them are hot, young
giant stars of the same type as Rigel,[1] telling us that the Pleiades
is an active and reasonably youthful association of stars which
has probably not been shining for more than perhaps 20-30
million years. Their distance is between 450 and 500 light-years
from us.

Close to the Pleiades in the sky, but considerably closer to us
in space, is the rather looser cluster of the Hyades, shaped like
a 'V' and seen by the ancients as the muzzle of Taurus, the
Bull. The bright orange star Aldebaran forms the eye of the
Bull but does not belong to the Hyades cluster as it is considerably closer to us. This group, which is about 135 light-years
away, is also a family, ranging from giants to dwarfs, but it
appears that many of its members are now past their prime. The
brilliant young giants are absent and although it is probable
that there is still much nebulosity in this group it is no longer
very apparent. Many of the stars in the Hyades have clearly
almost exhausted their supplies of fuel, and it appears to be a
much older cluster than the Pleiades.

Galactic clusters generally form local concentrations of stars
which can be easily identified as such, but some are much more
spread out. In their journey through space they gradually become separated and go their several ways. As often occurs in
families, the older they grow the more they become dispersed,
until eventually, after perhaps a thousand million years or so,
the family identity may be lost altogether. In the meanwhile
new families of stars are forever forming out of the gas and
dust of interstellar space, to become new clusters of stars in
the plane of the galaxy.

[1] See Page 140.

GLOBULAR CLUSTERS

Whilst a galactic cluster may occasionally contain as many as a thousand stars, a globular cluster may contain hundreds of thousands or even a million. As their name suggests, they are globular in shape, and it has been suggested that they resemble nothing so much as a swarm of bees clustered round their queen. The stars are very concentrated towards the centre and thin out rapidly towards the extremities, so that it is difficult to determine their full extent. The large clusters are probably almost a hundred light-years across.

Roughly a hundred globular clusters are known to belong to our galaxy, all of which are at tremendous distances from the sun and from the central plane. Only three, in fact, can be discerned with the naked eye, M13[1] in the constellation of Hercules, the only one in the northern sky (see Plate 19), with Omega Centauri and 47 Toucani conspicuous in the south.

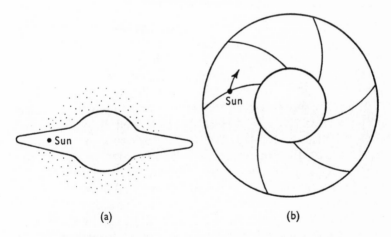

(a)   (b)

Fig. 17. A diagrammatic view of our galaxy seen (*a*) edgewise on, and (*b*) 'from the top'. The overall diameter is about 80,000 light-years, with the sun situated roughly 27,000 light-years from the centre. Note the distribution of the globular clusters which form an outer halo, mainly surrounding the central nucleus.

[1] 'M13' signifies the 13th object in Messier's catalogue of nebulous objects.

Many, however, are glorious objects when viewed with a telescope.

Globular clusters are rich in variable stars of the RR Lyrae[1] variety, so that their distances can be determined with a fair degree of certainty. Yet M13, which is clearly one of the closer clusters, is approximately 25,000 light-years from us. The fact that all save two of these objects were found to lie in one half of the sky led Shapley to suspect early in the 1920's that the sun was not in the centre of the galaxy as had been confidently believed, but was well removed to one side. He showed that these remarkable clusters are arranged as a sort of outer halo around the galaxy, being generally more concentrated around the central hub, as is indicated diagrammatically in Figure 17. This suggestion has been confirmed by the observation of globular clusters in the Andromeda galaxy and in other nearby spirals where the same arrangement is apparent.

STELLAR POPULATIONS

Once the distance of a globular cluster has been determined, no great inaccuracy is introduced if we assume all the stars in it to be at the same distance from us. Observation of the spectra of the individual stars thus enables a Hertzsprung-Russell diagram to be drawn for the stars of that cluster.

The surprising fact emerges that the stars in globular clusters do not fit into the normal Hertzsprung-Russell diagram that has been constructed for the stars in the regions near the sun. It now appears that there are two distinct *populations* of stars whose physical behaviours are somewhat at variance. Population I stars are of the type found in the spiral arms of galaxies (which includes the vicinity of the sun), where there is still an abundance of gas and dust. They fit neatly into the diagram as we have constructed it and their processes of formation appear to be still going on all around us. Population II are generally older stars which have long since out-grown the exuberance of their youth. They do not therefore include any of the blue-white giants with short lifetimes on the Main Sequence, but

[1] See Page 147.

seem to show what happens to stars later in their lives. They are found mainly in the globular clusters and in the central parts of our galaxy, both of which are regions apparently devoid of dust. Population II stars cannot, therefore, still be forming, except as aging products from stars of population I.

The significance of stellar populations will become more apparent when we look at the different types of galaxy, in the next chapter.

GASEOUS NEBULAE

Hidden away in the depths of space, and frequently only discernible with the aid of photography and a very large telescope, are the gaseous nebulae, some of which are among the most beautiful objects in the sky. The word 'nebula' simply means a 'mist' or a misty patch, and was at one time used to refer to any such object. It was applied in particular to the many millions of faint, hazy patches believed to lie beyond the limits of our own galaxy, but whose nature was still in doubt. These were called extra-galactic nebulae. As, however, we now know that these mysterious patches are all galaxies of one form or another, we will henceforward refer to them as such, and will confine our use of the term 'nebulae' to objects belonging to our own Milky Way system.

The gaseous nebulae occur in many different forms, of which the planetary nebulae are perhaps the most regular and 'starlike'. These are so called because when seen through a telescope they show a disc. No ordinary star can appear as anything more than a point of light even with the world's largest telescope, so it seems that these 'stars' must be enormous. It has been estimated that they probably measure a full light-year across and thus occupy a volume of space equivalent to about six hundred million solar systems. Their masses, however, appear to be no more than that of ten to fifteen suns, roughly the same as a super-giant, so they must obviously be very 'blown up' indeed. They consist of what, on earth, we would call an almost perfect vacuum! At the centre of each is a small but tremendously hot star with a surface temperature of the order of

100,000° C., which causes the whole globe of very rarefied gas surrounding it to fluoresce, and to display some beautiful colouring.

The Owl Nebula in Ursa Major (Plate 20) and the Ring Nebulae in Lyra are splendid examples of planetary nebulae which can be seen with telescopes of moderate power.

More common and of a very different type are the diffuse nebulae, of which the Great Nebula in Orion is the classic example (Plate 21). Situated as the central star of Orion's sword, it can be discerned with the naked eye simply as a hazy patch, yet here we have an immense, chaotic cloud of glowing gas about fifteen light-years across, shining simply as a result of the action of the very hot stars embedded in it. Like the Pleiades we might regard this as another wonderful example of nature at work, for here we clearly have stars in the process of being born. Within the cloud and visible through quite a small telescope are four stars[1] forming the Trapezium. These are certainly young and extremely hot, new-born blue-white stars, which cannot have been shining for more than ten million years. At present, the great cloud we are viewing has a density of only one thousand million millionth that of normal ' air '. Gradually, over millions of years, it will condense to form more stars. We may wonder when we look at its apparently wind-swept appearance whether, if we were to photograph the same region again in a year's time, we should notice any difference, but the answer is undoubtedly no. On our normal time-scale stellar evolution seems painfully slow. It is possible, however, that astronomers who photograph it perhaps 1,000 years from now may be able to detect a difference.

One thing that the radio astronomers can do for us which is of particular value is to detect the presence of neutral hydrogen gas. They have thus been able to establish that this wonderful nebula is simply one of the brighter regions of a vast cloud of inter-stellar gas which is spread over most of the region of Orion (Plate 22). As all this material is gradually condensing to form stars we can appreciate why it is that Orion possesses such a rich supply of brilliant objects.

[1] Careful examination with powerful instruments has revealed a total of eleven.

DARK NEBULAE

Unfortunately, however, not everything in the sky is beautifully illuminated for our observation. Dark clouds sometimes intervene and spoil the view. In our northern skies the Milky Way in the constellation of Cygnus shows many such clouds, and close at hand in Aquila it is divided by a great rift of darkness. Another very prominent dark patch appears near the Southern Cross and is known as ' the Coal Sack ' (Plate 23).

These objects, which appear as ' holes ' in the Milky Way, are in reality immense clouds of inter-stellar dust which lie between us and the brilliant stars beyond, blotting out their light. Occasionally stars can be seen against the darkness and appear as if shining through the hole (Gamma Cygni is an excellent example), but these stars are, in reality, closer to us than the cloud itself. The Coal Sack is about 400 light-years from us and probably contains the material for making somewhere between thirty and fifty suns. One day, perhaps, it will begin to take part in the gradual formation of new stars that appears to be going on all around us.

STARS IN THE MAKING

Stars are formed from the gas and dust of inter-stellar space. What starts them forming we do not know, but it seems certain that the process is going on the whole time and, as it were, is happening before our very eyes. Unfortunately the evolutionary time-scale is so great that it is impossible to observe any change in the span of a human lifetime. Nevertheless, if these processes are continuous, we should expect to find some stars which are just beginning their lives whilst for others their time is coming to an end (whatever that may mean). We should, of course, also expect to observe stars at all the stages in between. If we could arrange our observations in the correct order, should we then be able to produce the life history of a star?

Another question immediately arises. Do all stars have the same life history, or are there many stories spread out before us and jumbled together? If so, can we sort out which incidents belong to which story?

All these questions and many others of the same type have been concerning astronomers since the turn of the century. Without attempting to give all the reasons for their conclusions, let us now outline briefly how they believe the stars evolve.

In the first place it seems unlikely that stars are formed singly. The evidence from the galactic clusters we have studied indicates that they are probably born by families or associations numbering between 100 and 1,000. We do not yet know what causes the initial process of formation to begin, but it seems likely that once the necessary 'parcel' of dust and gas has come together, centres of concentration arise around which the material for making each star is gradually collected by gravitational attraction. The tenuous cloud of dust and gas that is to become a star then begins to collapse under the force of its own gravity, until eventually the pressure and temperature at its centre have risen to the tremendous values required to allow the thermo-nuclear processes to begin. As we saw in the case of the sun,[1] the internal pressure that now arises eventually counteracts any further collapse due to gravity and the star achieves a state of stable equilibrium. It has now taken its place on the Main Sequence on the Hertzsprung-Russell diagram (Figure 14) and the characteristic part of its life may be said to have begun.

Earlier in this chapter we noted that the hottest stars on the Main Sequence are the massive O and B type giants, and that luminosity falls off with mass as we move down the slope, past G type stars like our sun to the red dwarfs at the bottom of the scale. The position on the Main Sequence at which a star arrives thus depends primarily on the amount of matter it collected during its initial condensation. It is also this which governs the time it spends over forming. Massive stars condense relatively rapidly and their thermo-nuclear stages may be reached in a hundred thousand years or less. Small stars may take millions of years before they ever start to shine.

Once the Main Sequence is reached the star can settle down

[1] Page 81.

to use up its available supplies of fuel and it is in this stable state that it spends the main part of its life. How long this lasts depends on how long its fuel supplies hold out, and we have already seen how brilliant stars such as Rigel are spendthrifts who squander their substance at a fantastic rate. Their 'life' on the Main Sequence is thus unlikely to exceed ten million years. Stars such as our sun should be good for anything up to 40,000 million years, while at the lower end of the scale, where the stars are only glowing very feebly, the figure may be hundreds of times greater.

As the supply of hydrogen fuel at the centre of the star becomes exhausted gravitational contraction temporarily returns, so that further layers become heated and the thermo-nuclear processes gradually spread outwards. At the same time the contraction may raise the temperature of the central core to the region of a hundred million degrees so that as the process of converting hydrogen to helium stops, other nuclear reactions involving the conversion of helium and possibly other elements may start to take place. The star will begin to *grow* in size and, in effect, will leave the Main Sequence and progress towards the red giants. If it possesses planets this is the time when they cease to exist as they are gradually swallowed up by their parent. The time-scale for this process is not accurately known but is probably of the order of several cosmic years.[1]

However, the star has now virtually run down. It passes fairly rapidly across the Main Sequence again and, through suffering complete gravitational collapse, eventually finds its way to the old stars' home as a white dwarf. There its expenditure of energy is very small indeed and there is no knowing how long it may remain.

While astronomers in general would not claim that the intriguing sequence of events that we have just outlined is more than a hypothesis, it does appear to meet the observed facts and its details fall within the scope of the laws of nature as we know them. If we accept as a basis, some such sequence as this, we are in a reasonable position to speculate on the possibilities of there being any other life in such a universe of stars.

[1] 200,000,000 years, see page 164.

ARE WE UNIQUE?

At the end of Chapter III we considered the question of whether the solar system was the result of a naturally occurring process or whether it came into being through some unique accident. For the purposes of this speculation we shall assume the former to be the case, so that we can expect there to be millions of families of planets in the universe.

Let us now consider what conditions are necessary for the evolution of life on any one of them. In the first place it would obviously require that there should be a habitable zone around the star in question in which conditions should be neither too hot nor too cold for life to exist. Secondly, there must of course be a planet within that zone which has acquired a suitable atmosphere for supporting life. Finally these two conditions must be maintained for a sufficiently long period to enable life to evolve. We cannot, with certainty, say how long would be necessary but from our experience on earth it would appear that 1-2,000 million years would be a bare minimum.

From what we have seen of the life history of the stars it is clear that only the comparatively stable period spent on the Main Sequence is likely to produce these conditions. The hot stars at the upper end of the sequence, types O, B and A, are thus immediately ruled out because, although they would certainly possess broad habitable zones, their stable life would be far too short for the evolution of life. Cooler stars of type F and below appear more promising. All could offer the necessary stable period on the Main Sequence and all could produce habitable zones, although in the case of the very cool K and M stars these would be very narrow and close to the parent star.

The most favourable conditions for life therefore appear to exist in the vicinity of stars near the middle of the Main Sequence, those of spectral types F and G. Although approximately one star in every ten falls into this category, it seems likely that not more than half this number is eligible for consideration if we discount the double or multiple stars around which it would clearly be difficult for planets to ply a stable orbit. Thus it appears that about one star in every twenty might be capable of producing the necessary conditions for the evolution of life.

In a galaxy containing 100,000 million stars, this amounts to about 5,000 million. What, then, would be the chances of any one of them possessing a suitable planet on which life has actually evolved? Obviously this is a question to which we can only guess the answer, but if we were to put it at a million to one against, we would still be left with the possibility that there might be 5,000 inhabited planets in our galaxy. When we multiply this figure by perhaps a hundred thousand million for the number of galaxies in the universe, surely we are forced to admit that perhaps homo sapiens may not be as unique and important as many of us would have liked to think!

CHAPTER V

—••E)(3••—

THE GREAT UNIVERSE OF GALAXIES

. . . in space and time the universes,
All bound as is befitting each—
All surely going somewhere.
WALT WHITMAN

UNTIL BESSEL made the first measurement of the distance of a star in 1838 men could have had little conception of the real scale of our galaxy, let alone of the universe. Yet nearly half a century earlier William Herschel, while studying the thousands of nebulae revealed by the magnificent telescopes he had made with his own hands, became convinced that he was looking at what he called 'island universes'. He once made the profound statement: 'I have looked further into space than ever human being did before me. I have observed stars of which the light, it can be proved, must take two million years to reach the earth.' That he was correct in this belief was not finally proved for more than 120 years.

The fact that we live inside our galaxy, inevitably makes our attempts to learn about its structure very difficult. The problem might be likened to that of a small insect sitting on a clover leaf and trying to learn about the extent of the field in which it lives. Added to this there is, as we have seen, an immense amount of gas and dust in the plane of the galaxy which acts as an obscuring medium, effectively blocking our view of the more distant regions. It is as if the insect's field were shrouded in an early morning mist. Let us now imagine that the field was on the side of a mountain from which the insect could see, spread out before him in the valley below, fields of all shapes and sizes. From the little he could see of his own field, he might possibly be able to tell which of them it most resembled. In just the same way our researchers are greatly assisted by the fact that at a distance we are able to observe other galaxies, some of which are believed to be similar to our own, and so we can obtain a perspective view of how our galaxy itself might appear.

In our opening chapter we considered our galaxy scaled down to the size of a bun ten miles across and rather less than two miles thick, in company with millions of other buns which made up the universe. Since then we have learned something of the ingredients of which this bun is made and a little about the arrangement of its crumbs. We have seen how the Milky Way is the result of the effect of depth on our vision, how the galactic clusters are formed from the gas and dust which pervade the central plane, how the sun is situated roughly two-thirds of the distance out from the centre, how the globular clusters are distributed in an outer haze, mainly around the central hub, how the stars appear to be divided into Population I and Population II, whose physical processes are different and who occupy different regions of the galaxy. We have also stated that our galaxy has spiral arms and that it is in rotation, although we have not so far in any way justified these last statements.

Figure 17 shows diagrammatically what we believe our galaxy to look like. Latest estimates put it at about 80,000 light-years across, so placing the sun at a distance of roughly 27,000 light-years from the centre. It is believed to be largely held together by the gravitational attraction of the central nucleus, behaving rather like the sun at the centre of his family of planets. If the galaxy were in rotation we should not, therefore, expect it to turn like an enormous cartwheel, with every part completing one revolution in precisely the same period of time, but rather we should expect to find those portions nearest the centre moving considerably faster than those near the extremities, just as in the solar system Mercury dashes round close to the sun while Pluto patrols sedately around the outer perimeter. There should thus be a noticeable shearing effect between any one group of stars and the next group inwards or outwards in the galaxy.

This effect has indeed been observed. By studying the bright O and B type stars at great distances from us in the galactic plane, we can find the differences in their average speeds and thus determine that the stars in the vicinity of the sun are moving round in the galaxy with a speed of about 160 miles per second. From this it can be calculated that our sun takes about 200 million years to complete one revolution of the galaxy, an interval of time which is known as a cosmic year. At the same time it can be calculated from the forces at work that the galaxy contains the necessary material to make about a hundred thousand million stars the size of our sun.

The behaviour of other galaxies strongly suggests to us that our own Milky Way system is a spiral, although proof has been difficult because of our own location inside the system. Ambitious attempts to plot the direction and distance of vast numbers of stars, with the hope of revealing a spiral pattern, were inconclusive. Finally, the radio astronomers took a hand and their efforts have led to a definite mapping of the system. The advantage came from the fact that the vast clouds of dust, which occult the visible structure of the distant spiral arms, are transparent to the radio waves of neutral hydrogen on its wavelength of 21 centimetres. In this way the spiral structure of the Milky Way has been definitely demonstrated.

Radio telescopes thus provide a wonderful means of tracing out the distribution of hydrogen, which is by far the most common element in the universe, and these investigations have already shown up the spiral structure of our galaxy in a most convincing manner and beyond any doubt. A structural map of the galaxy is thus being built up and further researches in this field based on a combination of both radio and optical astronomy appear to hold out great prospects for the future.

VARIETIES OF GALAXY

Although we have frequently likened the galaxies to the buns in our model, we were careful to note in Chapter I that in reality they differed widely in both shape and size. Galaxies, like stars, appear to exist in a very wide variety of different types. Some are perfectly symmetrical, others are without any apparent form; some are globular in shape, others appear extremely flat. As with the stars we must endeavour, by examining all their characteristics, to arrange the various specimens into a sequence which might tell the story of their lives.

The galaxies can be divided into three main types; irregular, spiral and elliptical, and with some modifications it seems probable that this order represents their evolutionary sequence. Specimens of each type are readily available for observation from our own galaxy, so we will examine each in turn.

IRREGULAR GALAXIES

In the southern sky, not far from the South Celestial pole, are the two irregular galaxies known as the Magellanic Clouds (Plate 16). Originally these were known as the ' Cape ' Clouds, because those intrepid Portuguese seamen used to keep them ahead of their ships as they pressed south in their quest for a route round the Cape of Good Hope. Later they were named after Magellan, the circumnavigator of the globe. Although these two clouds give the appearance of being detached fragments of the Milky Way, they are in reality far removed from our own system, the Small Magellanic Cloud and the Large Magellanic Cloud lying at distances of approximately 100,000 light-years. Even at this great distance they are our nearest neighbours in extra-galactic space. In effect they may even be connected to the Milky Way system through gravitational attraction and may well share its general motion through space. Since both of these clouds are smaller than our own galaxy we might liken them to the satellite towns of our great star city.

The comparative closeness of the Magellanic Clouds makes them of tremendous importance to us, for it is possible to identify specimens of all the principal interesting objects we encountered in our own galaxy : globular clusters, galactic clusters, gaseous nebulae, super-giants, Cepheid variables, and so on. All these can be studied in perspective and true comparisons can be made, for all can be regarded as being at the same distance. The clouds are also extremely prominent in the 'radio sky', and study of their twenty-one centimetre radiation has shown that both are embedded in immense clouds of hydrogen, so extensive that they may very well be in contact with one another and possibly even with our own galaxy.

It seems certain that in the Magellanic Clouds we have specimens of comparatively young galaxies, although what may be meant by 'young' in this context is still to be determined. Perhaps, for the time being, we should describe them simply as 'early developments'. They have very little established form and they contain predominantly the younger Population I stars, so that we can, with some confidence, place them somewhere near the beginning of the life story of the galaxies. Nevertheless, we should not jump too quickly to the conclusion that they are necessarily younger in 'age' than other galaxies. There may well be other factors than the mere passage of time which determine the rate of development of these interesting systems.

SPIRAL GALAXIES

Undoubtedly the most impressive variety of galaxy is the spiral, of which we have two striking examples in M31 and M33[1] within easy 'observable' reach (Plates 25 and 26). M31 is the Great Andromeda galaxy, which on a clear night can be picked out with the naked eye as a hazy patch of light of about the fifth magnitude. M33 is in much the same part of the sky, in the constellation of Triangulum, but is just beyond the threshold of normal vision. Their distance is of the order of two million light-years, with the Andromeda Spiral slightly the closer.

When we look at a photograph such as that in Plate 25, and talk casually about these tremendous distances, it is as well, from time to time, to remind ourselves of what they really mean. About two million years ago, long before men existed on earth, a ray of light started on its journey from the Andromeda galaxy towards the earth, travelling across 186,000 miles of space every single second. After about a million years, when it

[1] See Footnote on Page 153.

had covered roughly half the journey, 'men' in their most primitive form began to appear. Almost a million years later still, these men began to become civilised, and amongst other activities, to study astronomy. All this time the ray had been approaching at a speed of 186,000 miles per second. Just in the last fraction of its journey man invented a telescope and by pointing it in the right direction was able to catch that ray as it arrived and so take the wonderful photograph we now possess. To see this galaxy as it is to-day we should of course have to wait patiently for another million and a half years. M31 is one of the nearer galaxies. When photographing the most distant we may be catching rays of light which have been travelling since the earth itself was young, and possibly since before the earth even existed!

Both M31 and M33 show a bright nucleus and a well-defined spiral structure, although the latter is clearly in a somewhat earlier stage of development. The distinctive pin-wheel appearance of both galaxies suggests a definite rotation and this has been confirmed by spectroscopic observations of the upper and lower limbs which clearly indicate that one part is approaching us while the other is receding. In M31 it has also been possible to note the different speeds of rotation at different distances from the nucleus which we have already remarked on in our own galaxy and the direction of rotation has been shown to be that which would tend to wind up the spiral arms more tightly.

It was, in fact, whilst studying M31 that Baade of the United States first observed the distinctive distribution of the Population I and II stars. He noticed how the brilliant young Population I blue-white giants were found only in the outer regions of the galaxy where they are embedded in masses of gas and dust, while the central core, which appears not to take part in the general galactic rotation, consists predominantly of elderly, cool, red stars of Population II. More recently it has become possible to photograph this galaxy on colour film, and the same distribution of colour is clearly visible.

The Andromeda galaxy is one of the biggest galaxies that are reasonably close to us, and is believed to be considerably larger than our own, probably measuring some 120,000 light-years across. It is interesting to note, however, that until 1952 it was thought to be at a distance of only about three-quarters of a million light-years, roughly half our present estimate. When, in 1952, Baade obtained a true calibration for the Cepheid vari-ables,[1] he virtually doubled all extra-galactic distances, at the same time doubling the accepted size of the universe! In con-sequence it became apparent that M31 must really be double the size that had previously been accepted, and thus consider-ably larger than our own system.

Plates 27 and 28 show other spiral galaxies viewed from different angles, one ' full face ' and the other edgewise on. Both illustrate clearly the typical shape and beautiful structure of this variety. The outer regions of N.G.C.4565 in Plate 28 give evidence of considerable inter-stellar absorption, indicating the presence of gas and dust within which the birth of stars is clearly still proceeding.

Spiral galaxies occur in many different forms, in some of which the spiral structure is only just becoming evident whilst others are already tightly wound up. Some, again, are known as barred spirals because the spiral arms appear to spring from the end of a central ' bar '. In either case it is generally believed that the tightest wound spirals are the most advanced in the evolu-tionary sequence, and for this reason the spirals are sometimes classified as early or late according to their degree of develop-ment.

ELLIPTICAL GALAXIES

In Plate 25 it will be observed that the Andromeda spiral is accompanied by two smaller galaxies, both elliptical in shape, which are clearly satellite galaxies associated with the main system, rather as the Magellanic Clouds may be connected to our own. Other elliptical systems have been found which are enormous, and comparable in size with the largest known spirals. Their shape varies considerably from practically spherical to

[1] Page 145.

highly elliptical, yet their general characteristics are otherwise remarkably uniform. They show very little structural form yet are invariably symmetrical, they consist almost entirely of old stars of Population II and, as they appear to possess no obscuring inter-stellar dust, they are frequently transparent so that more distant galaxies can be observed right through them. Many are accompanied by haloes of globular clusters.

In the elliptical galaxies we seem once more to have reached the end of the evolutionary sequence which started with the irregular galaxies and developed through the spirals, in various stages of winding up, to its apparently logical conclusion. Once again, however, as in the case of the stars, this is mainly a hypothesis which appears to meet the facts, and we should avoid saying that elliptical galaxies are necessarily older in time than the others.

CLUSTERS OF GALAXIES

Just as we found that the stars form into clusters, so apparently do the galaxies. Although they are separated by fantastic distances, it appears that they are grouped together into associations which are separated by even greater distances. As with the stars, the sizes of these clusters vary considerably from a mere handful up to enormous groups with many hundreds of members.

Our own galaxy belongs to a fairly small cluster which we know as the local group and which contains only seventeen members, of which three are spirals (our own, M31 and M33), four are irregulars and no less than ten are elliptical. It appears therefore, that our immediate neighbours consist mainly of old age pensioners in the cosmological sense. Some of the more distant clusters, however, appear to contain a much higher proportion of spirals. Although the significance of this grouping is not yet fully understood it appears to be more than a mere accident. Each group shares a common movement through space, although within the groups themselves there is a fair amount of random movement. It is as if the whole population of galaxies were divided amongst a number of reasonably adjacent rafts. The rafts can move bodily relative to one another, whilst on each raft the individual members can move about as they please. Such an organisation extends from our own local system to the limits of our observational horizon, which with the 200-inch telescope on Mount Palomar may be taken to be at a distance in excess of 2,000 million light-years. As we saw in Chapter IV, it is probable that developments in the field of photo-electric photometry may well extend this limit, and already there have been isolated instances of galaxies being photographed at distances more than double this figure. At present, however, these are rare exceptions. It has been estimated that there are roughly one hundred million galaxies within the range of the Palomar Giant, and the most distant known cluster lies at about 2,000 million light-years, in the constellation of Hydra.

What happens beyond this limit? Every increase in telescopic power has revealed more galaxies at greater distances, and the radio astronomers, as we have seen, confidently believe that they have already probed to several times this distance. Does the universe go on for ever?

Before we answer that question there is one other observed feature of the galaxies which we should discuss.

THE EXPANDING UNIVERSE

Perhaps the most far-reaching cosmological discovery that has been made this century has been that of the red-shift in the light reaching us from the galaxies. By now we are familiar with what this phenomenon means when observed in the light coming from a star, and are prepared to accept without question that the body is receding from us. Yet when we apply this to the universe as a whole it appears that no matter in which direction we look, the groups of galaxies are moving away from us, and in some cases with enormous speeds. The whole universe seems to be expanding, and the further away *from us* a galaxy is, the faster it *appears* to be receding. At a distance of 2,000 million light-years galaxies seem to be moving away with speeds of about 37,000 miles per second, which is one-fifth of the speed of light. Yet it is all very well to say the whole of space appears to be expanding, but what can such a statement really mean?

We are tempted to ask, what is it expanding into, or does space go on for ever?

This is the point where our everyday conception of space and Newton's theory of gravitation both break down, and Einstein and his colleagues take over. Space has been described as boundless yet finite, and although this remarkable statement seems to be a contradiction in terms, a simple illustration may help to show what is meant.

Let us imagine a microbe confined to the surface of a spotted balloon. He is only capable of appreciating two dimensions and so his world is apparently flat. He can move wherever he wishes over the surface of the balloon and he will never come to any boundary. It appears to him that his world is boundless, yet to us it is undoubtedly finite.

Now let us imagine the balloon being blown up. To the microbe the surface begins to expand. No matter in which direction he looks, the spots on the surface seem to be moving *away* from him, just as the galaxies appear to be moving away from us. At the same time they are, of course, also moving away from one another.

As this simple example was explained in terms of two dimensions, we were able to understand its significance because our everyday experience enabled us to appreciate the third dimension of which the microbe was not aware.

Space can be likened to the surface of the balloon, and can thus be considered as curved and expanding in just the same way. Notice, however, that in this comparison it is the *surface* of the balloon and not the balloon itself which is being considered. Unfortunately, while we can understand a curved two-dimensional surface, we cannot imagine what a curved three-dimensional space would be like in terms of our everyday experience. Nevertheless, mathematically such an idea is perfectly acceptable and we can talk about curved space, and a boundless and finite universe, without being in any way contradictory.

As the galaxies at a distance of 2,000 million light-years appear to be receding from us at about one-fifth of the speed of light, we may be tempted to ask, ' If we could see five times as far as we can now, should we then have reached the limit? ' At that distance we would expect the galaxies to be receding at the speed of light and this would appear to mean that their light could never reach us. We should then have reached the limit of the observable universe. This we certainly believe to be the case, but at present such questions cannot be answered with complete confidence. Already, however, we have seen sufficient to realise that even the vast volume of space which we can explore can be nothing but a minute fraction of the whole great cosmos.

Sir Arthur Eddington once estimated that in just the same way as we say there are about a hundred thousand million stars in an average galaxy, he considered there might be a hundred thousand million galaxies in the universe. This gives a figure of 10 sextillion stars.[1] It is a reasonable figure to work on, and begins to make our little earth look pretty insignificant by comparison. Our petty troubles and squabbles no longer seem to matter when considered against such a background. Perhaps the human race would achieve a better sense of proportion and learn to become more tolerant if it were to think of things in these terms rather more often.

HOW DID IT ALL BEGIN?

We have now seen how observational astronomy tells us about the universe out to a distance of about 2,000 million light-years, thus enabling us to see portions of it as they existed 2,000 million years ago. If we could see further into space this would of course be the equivalent of seeing further back in time. The possibilities of doing this are unfortunately severely limited. Apart from the physical difficulties of building ever larger telescopes (and the Russians are already well advanced with their 236-inch reflector) there is the fact that if the galaxies really are receding from us, less and less of the light they emit will actually ever reach our instruments.

The possible reasons for this apparent expansion of the universe are beyond the scope of this book, but if it is accepted that this is the true explanation of the red-shift, and the majority

[1] Ten thousand million million million (10^{22}).

of leading cosmographers are of the opinion that it must be, then about ten thousand million light-years is the probable limit open to observation. It is thus unlikely that even with the aid of radio astronomy our present horizon can be extended by more than perhaps a few thousand million light-years, and whatever lies beyond that distance has already gone out of our range for ever. However, although it is doubtful if we shall ever be able to see far enough into the past to see the beginning, we may still be able to see sufficiently far to say with some degree of certainty how it all began.

At the present time the rival cosmological theories can be divided into two main groups which have become known as the Evolutionary and the Steady State theories. Each has its adherents among the advanced thinkers of our age, and we shall briefly consider each in turn.

If it is true that the galaxies are indeed racing away from us, and that the further they are the faster they are moving, it follows that a million years ago they must have been very much closer to us, and to each other, than they are to-day. A million years before that they would have been closer still. If, therefore, we could go on putting the clock back we should eventually find a time when the galaxies were all concentrated together. Calculation has shown that, if our measuring rods are correct, this would have occurred between eight and nine thousand million years ago. Le Maitre, whose name has been associated with a number of evolutionary theories, envisaged a time when all the matter in the universe was concentrated together into a conglomerate of unbelievably high density, which has been called the *primeval atom*. Some event, we cannot tell what, caused this conglomerate to disrupt, so that it began to expand. Thousands of millions of years went by and the material eventually reached a state approaching stable equilibrium again, in which the tendency to expand further was countered by the force of gravitational attraction. This tendency to expand is something which at present is not fully understood, although certain theories exist which attempt to explain it. In Einstein's universe it is simply a mathematical term which has been called the 'force of cosmical repulsion'.

With the universe once more in a stable state the galaxies gradually began to form from the material of the primeval atom. No timetable can be given for this period but it must have lasted a very long time, probably many thousands of millions of years. Eventually, and possibly as a result of the formation of the galaxies, the force of cosmical repulsion again became the stronger, expansion started once more, and after some nine thousand million years the galaxies became spread out in the way in which we know them to-day.

Many variations of this evolutionary theory have been proposed by others. Gamov has suggested that the force of the initial explosion was sufficient in itself to produce the expansion we are now witnessing. Another suggestion is that the primeval atom was simply the stage of maximum contraction of a universe that may have been in existence since eternity. Generally speaking, however, all evolutionary theories are based on the idea of a single act of creation, which science cannot at present attempt to explain.

We should be careful to note that, in each case, we come to a state of affairs beyond which we are unable to proceed. We have not, of course, arrived at the beginning of the universe, but simply at a point beyond which science can no longer guide us.

Before we examine the evidence in support of this group of theories, let us also look at the alternative group which have become known as the Steady State theories. These were first suggested by Hoyle, Bondi and Gold at Cambridge in 1948 and have received a fair measure of support from astronomers throughout the world.

In the evolutionary theories the galaxies are forever receding from our view beyond the observable horizon. It follows, therefore, that the number of galaxies still within the limits open to our exploration must be for ever decreasing, for there is nothing to take the place of those that have gone for ever beyond the range of our telescopes. In the Steady-State theory, however, new galaxies are presumed to be coming into existence all the time to fill in the gaps in the expanding universe and, as it were, to take the place of those that have drifted out of our view. They are, in fact, being formed all the time.

Unfortunately, this Steady-State or constant-density theory for the evolution of the universe has sometimes popularly been referred to as ' continuous creation '. As a consequence, the theory has attained almost a religious connotation. It has been said that if God intended to create a universe, why should He do it in one creative act and then stop? Instead of producing the universe in one big bang, why should He not ' create ' it in innumerable tiny ' pops '?

Let us see whether, within our limited capabilities, we can put these conflicting theories to the test. If the Steady-State theory is correct, matter must be appearing spontaneously throughout the universe at this moment. Is it possible for us to detect this matter in its initial state or later in its life as it starts to form into galaxies?

The volume of space is so vast that even the fantastic quantity of matter that would be required to replace the vanishing galaxies would only amount to the equivalent of injecting a few atoms of newly-created hydrogen into a space the size of the Empire State building every million years. It seems unlikely that such a small quantity will be detectable in the foreseeable future. On the other hand serious claims have been put forward that certain galaxies which have been named ' peculiar ' galaxies and are as yet ' unformed ', are in reality very young galaxies that have recently started to come into existence. Some are comparatively close to us, which means that we must be seeing them as they were not so very long ago. If they really are young, by which we mean they are still in the process of forming from the primitive material of space, they cannot have been in existence for 8-9,000 million years.

Surely here is a pointer towards the truth of the Steady-State theory. It is not conclusive, for there have been other explanations to cover the case of arrested development amongst the galaxies, but it is certainly significant.

Turning to the evolutionary theories, we have already seen that if the universe had a beginning, say 9,000 million years ago, then the conditions that existed then would be very different from those we observe to-day. In particular the galaxies would be very much closer together than they now appear. By the

Steady-State theory, on the other hand, 9,000 million years ago things would appear precisely the same as they do now. Although it is not possible for us to see as far backwards in time as this, it is apparent that if we could see even 4-6,000 million years into the past the galaxies should still be considerably closer together than we observe them in our vicinity at the present time.

So far attempts to verify this by means of optical astronomy have failed, for our horizon is too limited and we are unable to see far enough back in time. Here, again, the radio astronomer comes to our aid.

Except in a few isolated cases, of which the sun is an example, radio sources cannot in general be equated with individual stars. Those sources which belong within our galaxy owe their origin to a number of different causes including, as we have already seen, such explosions as resulted in the Crab Nebula.[1] Others have been proved to originate beyond the limits of our galaxy, and the most powerful radio source in the sky has been identified with a pair of colliding galaxies in the constellation of Cygnus which are believed to be at a distance of 500 million light-years. It is only very powerful radio sources of this nature which are likely to be detectable out to the fantastic distances of 4-6,000 million light-years which are required to provide the proof for which we are searching. The density of such sources in our vicinity can, of course, be established. However, experiments carried out at the Radio Observatory at Cambridge under Professor Ryle have shown that at greater distances this density is markedly increased. The number of detectable sources at distances between three and eight thousand million light-years appears to be at least three times, and in some cases ten times, what would be predicted by the steady state theory.

Brilliant as these findings undoubtedly are, we must avoid the temptation to jump to unjustified conclusions. Even this powerful piece of evidence is, at best, only another pointer in favour of some form of evolutionary theory. It may seem to indicate that the universe started with a ' bang ', but it does not in any way help us to choose between the various evolutionary

[1] Page 150.

models that have been proposed, neither does it yet rule out the idea of the continuous creation of matter. Some form of modified steady state theory is indeed still possible.

There are, of course, many other lines of evidence that can be produced in support of one theory or the other, but a full appraisal of this see-saw controversy is outside the scope of this book. The truly significant factor that emerges from recent trends in observational astronomy is that during the past decade a change has come over our whole mode of thinking about these matters. No longer does cosmology lie purely in the realms of philosophy; it has become a definite science capable of providing proof of its theories. Despite its already impressive contributions to our knowledge the wonderful tool of radio astronomy must be regarded as being purely in its infancy, comparable perhaps with optical astronomy a decade or two after Galileo made his first classic pronouncements. The next few decades will see both optical and radio observatories established in space, so that the complete electro-magnetic spectrum will at last be open to our investigation.

We can already see the pattern of new and exciting discoveries. Astronomers have discovered a new variety of celestial object, known as the ' Quasar ', or quasi-stellar radio source. These objects reveal themselves by being sources of vast amounts of radio energy. Specially built radio telescopes locate their position with high precision. Optical examination of these areas reveals nothing very startling in the way of astronomical objects except, in a few cases, of slightly fuzzy patches or ' stars ' that vary irregularly in brightness. Various theories have been suggested concerning the nature of the Quasars. The most appealing identify them as ' super stars ', objects having the mass of perhaps a million ordinary stars. Although such stars should normally be expected to collapse and explode, it may well be that intense magnetic fields stabilise the Quasar and control its explosive character. Another type of radio source, possibly remotely related to the Quasar, is the so-called ' blue galaxy ', which appears to be a distant object of exceptional blueness and therefore of extremely high temperature. The discovery of such remarkable objects indicates that astronomers still have much to learn.

Late in 1965, Professor Fred Hoyle of Cambridge, England, one of the original proponents of the Steady-State theory, surprised the scientific world by announcing that he had given up that theory. The mere existence of quasars, he conceded, disproved the concept that the universe appears the same, wherever or whenever one views it. For quasars appear to exist in the most distant and presumably the youngest galaxies visible to us. Various astronomers still hold to the Steady-State theory.

And, as we learn more of the wonderful mysteries of the universe who can say that our added knowledge will not of itself raise questions the like of which are at present still beyond our comprehension? Galileo could surely never have wondered about the thermo-nuclear processes going on inside a star. As we top the next mound, instead of finding the level plain of understanding spreading clear before us, may we not see a whole new mountain range in the distance, beckoning us onwards?

APPENDIX I: PLANETARY DATA

		Mercury	Venus	Earth	Mars	Jupiter	Saturn	Uranus	Neptune	Pluto	Sun
Mean Distance from Sun (Millions of miles)		36·0	67·2	92·9	141·5	483·3	886	1782	2792	3649	—
Mean Distance from Sun (Astronomical Units)		0·39	0·72	1·0	1·52	5·20	9·54	19·18	30·03	39·28	—
Sidereal Period (years)		0·24	0·62	1·0	1·88	11·86	29·46	84·01	164·8	248·4	—
Inclination of Orbit to Ecliptic (degrees)		7·0	3·4	0	1·9	1·3	2·5	0·8	1·8	17·2	—
Inclination of Equator to Orbit (degrees)		?	32 ?	23·45	24·0	3·1	26·75	97·9	28·8	?	7·25
Mean Orbital Velocity (m.p.s.)		29·7	21·7	18·47	15·0	8·1	6·0	4·2	3·4	3·0	—
Diameter (miles)		3,100	7,700	7,927 7,900	4,216	88,700 82,800	75,100 67,200	29,300 32,000	27,700	4,900	864,000
Volume (x Earth)		0·06	0·9	1·0	0·15	1,312	763	50	43	0·2 ?	1,300,000
Mass (x Earth)		0·06	0·81	1·0	0·11	318	95·2	14·6	17·3	0·1 ?	333,434
Density (x Water)		5·1	4·97	5·52	3·96	1·34	0·70	1·27	2·2	?	1·41
Surface Gravity (x Earth)		0·36	0·87	1·0	0·38	2·64	1·17	0·92	1·4 ?	?	28·0
Velocity of Escape (m.p.s.)		2·6	6·4	6·9	3·1	37	22	14	15	?	384
Rotational Period		88 days	?	23 hrs. 56 m.	24hrs. 37 m.	9h.50·5 9h.55·7	Eq. 10h.14	10h.49	14h. ?	6·39 d.	25·38 d.
Max. Surface Temp.	F.	770	140	140 ?	86	−216	−243	−300	−330	−348	10,800
	C.	410	60	60 ?	30	−138	−153	−185	−200	−211	6,000
Albedo		0·07	0·59	? 0·5 ?	0·15	0·44	0·42	0·45 ?	0·52 ?	?	—
Number of Moons		0	0	1	2	12	9	5	2	?	—

APPENDIX II: THE CONSTELLATIONS

Name of Constellation	Genitive	Translation
Andromeda	Andromedae	Andromeda
Antlia	Antliae	Pump
Apus	Apodis	Bird of Paradise
Aquarius	Aquarii	Water Carrier
Aquila	Aquilae	Eagle
Ara	Arae	Altar
Argo[1] (See Carina, Vela and Puppis)		Ship
Aries	Arietis	Ram
Auriga	Aurigae	Charioteer
Bootes	Bootis	Herdsman
Caelum	Caeli	Chisel
Camelopardus	Camelopardi	Giraffe
Cancer	Cancri	Crab
Canes Venatici	Canum Venaticorum	Hunting Dogs
Canis Major	Canis Majoris	Greater Dog
Canis Minor	Canis Minoris	Lesser Dog
Capricornus	Capricorni	Sea Goat
Carina	Carinae	Keel
Cassiopeia	Cassiopeiae	Cassiopeia
Centaurus	Centauri	Centaur
Cepheus	Cephei	Cepheus
Cetus	Ceti	Whale
Chamaeleon	Chamaeleontis	Chameleon
Circinus	Circini	Compasses
Columba	Columbae	Dove
Coma Berenices	Comae Berenices	Hair of Berenice
Corona Australis	Coronae Australis	Southern Crown
Corona Borealis	Coronae Borealis	Northern Crown
Corvus	Corvi	Crow
Crater	Crateris	Bowl
Crux	Crucis	Southern Cross
Cygnus	Cygni	Swan
Delphinus	Delphini	Dolphin
Dorado	Doradus	Swordfish
Draco	Draconis	Dragon
Equuleus	Equulei	Easel
Eridanus	Eridani	River
Fornax	Fornacis	Furnace
Gemini	Geminorum	Twins
Grus	Gruis	Crane
Hercules	Herculis	Hercules
Horologium	Horologii	Clock
Hydra	Hydrae	Hydra
Hydrus	Hydri	Water-snake

[1] The constellation of Argo Navis is now divided into the separate constellations of Carina, Vela and Puppis. One sequence of Greek letters is used through the three constellations.

Name of Constellation	Genitive	Translation
Indus	Indi	Indian
Lacerta	Lacertae	Lizard
Leo	Leonis	Lion
Leo Minor	Leonis Minoris	Lesser Lion
Lepus	Leporis	Hare
Libra	Librae	Scales
Lupus	Lupi	Wolf
Lynx	Lyncis	Lynx
Lyra	Lyrae	Lyre
Mensa	Mensae	Table (Mountain)
Microscopium	Microscopii	Microscope
Monoceros	Monocerotis	Unicorn
Musca	Muscae	Fly
Norma	Normae	Square
Octans	Octantis	Octant
Ophiuchus	Ophiuchi	Ophiuchus
Orion	Orionis	Orion (Hunter)
Pavo	Pavonis	Peacock
Pegasus	Pegasi	Pegasus (winged horse)
Perseus	Persei	Perseus
Phoenix	Phoenicis	Phoenix
Pictor	Pictoris	Painter
Pisces	Piscium	Fishes
Piscis Austrinus	Piscis Austrini	Southern Fish
Puppis	Puppis	Poop
Pyxis	Pyxidis	Compass (Mariners)
Reticulum	Reticuli	Net
Sagitta	Sagittae	Arrow
Sagittarius	Sagittarii	Archer
Scorpio	Scorpii	Scorpion
Sculptor	Sculptoris	Sculptor
Scutum	Scuti	Shield
Serpens	Serpentis	Serpent
Sextans	Sextantis	Sextant
Taurus	Tauri	Bull
Telescopium	Telescopii	Telescope
Triangulum	Trianguli	Triangle
Triangulum Australe	Trianguli Australis	Southern Triangle
Tucana	Tucanae	Toucan
Ursa Major	Ursae Majoris	Great Bear
Ursa Minor	Ursae Minoris	Little Bear
Vela	Velorum	Sails
Virgo	Virginis	Virgin
Volans	Volantis	Flying Fish
Vulpecula	Vulpeculae	Fox

INDEX

—••E)(3••—

183

Tennyson, 151
Terminator, 66, 68
Thales of Miletus, 18
Tides, 70, 71
Titan, 111
Trapezium, 156
Tropics, 56, 57
Troposhere, 43
Twins (Constellation), 18, 58, 156
Tycho Brahe, 21, 28, 74, 149
Tycho (Crater), 69, 74

Ungava, Lake, 75
Universe, 35-37, 170-177
Uranus, 32, 112, 113
Ussher, James, 63

Van Allen belts, 51, 108

Variable stars, 144-148
Vasco da Gama, 20
Vega, 62, 130, 131, 137, 138
Velocity of escape, 80
Venus, 23, 31, 92, 93, 96-99,
    Plate 9
Vesta, 116
Virgin, 126
Von Weizsacker, 122
Vulcan, 95, 96

Watts, Isaac, 126
White dwarfs, 141, 142, 159
Wolf-Rayett stars, 133

X-rays, 44, 45

Zodiac, 18, 58, 126

CATALOGUE OF DOVER BOOKS

ASTRONOMY

THE INTERNAL CONSTITUTION OF THE STARS, Sir A. S. Eddington. Influence of this has been enormous; first detailed exposition of theory of radiative equilibrium for stellar interiors, of all available evidence for existence of diffuse matter in interstellar space. Studies quantum theory, polytropic gas spheres, mass-luminosity relations, variable stars, etc. Discussions of equations paralleled with informal exposition of intimate relationship of astrophysics with great discoveries in atomic physics, radiation. Introduction. Appendix. Index. 421pp. 5⅜ x 8.
S563 Paperbound **$2.75**

PLANETARY THEORY, E. W. Brown and C. A. Shook. Provides a clear presentation of basic methods for calculating planetary orbits for today's astronomer. Begins with a careful exposition of specialized mathematical topics essential for handling perturbation theory and then goes on to indicate how most of the previous methods reduce ultimately to two general calculation methods: obtaining expressions either for the coordinates of planetary positions or for the elements which determine the perturbed paths. An example of each is given and worked in detail. Corrected edition. Preface. Appendix. Index. xii + 302pp. 5⅜ x 8½.
S1133 Paperbound **$2.25**

CANON OF ECLIPSES (CANON DER FINSTERNISSE), Prof. Theodor Ritter von Oppolzer. Since its original publication in 1887, this has been the standard reference and the most extensive single volume of data on the calculation of solar and lunar eclipses, past and future. A comprehensive introduction gives a full explanation of the use of the tables for the calculations of the exact dates of eclipses, etc. Data furnished for the calculation of 8,000 solar and 5,200 lunar eclipses, going back as far as 1200 B.C. and giving predictions up to the year 2161. Information is also given for partial and ring eclipses. All calculations based on Universal (Greenwich) Time. An unsurpassed reference work for astronomers, scientists engaged in space research and developments, historians, etc. Unabridged republication, with corrections. Preface to this edition by Donald Menzel and Owen Gingerich of the Harvard College Observatory. Translated by Owen Gingerich. 160 charts. lxx + 538pp. 8⅜ x 11¼.
S114 Clothbound **$10.00**

THEORY OF THE MOTION OF THE HEAVENLY BODIES MOVING ABOUT THE SUN IN CONIC SECTIONS, Karl Friedrich Gauss. A landmark of theoretical astronomy by the great German scientist. Still authoritative and invaluable to the practicing astronomer. Part I develops the relations between the quantities on which the motion about the sun of the heavenly bodies depends—relations pertaining simply to position in the orbit, simply to position in space, between several places in orbit, and between several places in space. The calculation methods of Part II based on the groundwork of Part I include: determination of an orbit from 3 complete observations, from 4 observations (of which only two are complete), determination of an orbit satisfying as nearly as possible any number of observations whatever, and determination of orbits, taking into account the perturbations. Translation of "Theoria Motus" and with an appendix by C. H. Davis. Unabridged republication. Appendices and tables. 13 figures. xviii + 376pp. 6½ x 9¼.
S1056 Paperbound **$2.95**

STAR NAMES AND THEIR MEANINGS, Richard Hinckley Allen. An unusual book documenting the various attributions of names to the individual stars over the centuries. Here is a treasure-house of information on a topic not normally delved into even by professional astronomers; provides a fascinating background to the stars in folk-lore, literary references, ancient writings, star catalogs and maps over the centuries. Constellation-by-constellation analysis covers hundreds of stars and other asterisms, including the Pleiades, Hyades, Andromedan Nebula, etc. Introduction. Indices. List of authors and authorities. xx + 563pp. 5⅜ x 8½.
T1079 Paperbound **$2.35**

A HISTORY OF ASTRONOMY FROM THALES TO KEPLER, J. L. E. Dreyer. (Formerly A HISTORY OF PLANETARY SYSTEMS FROM THALES TO KEPLER.) This is the only work in English to give the complete history of man's cosmological views from prehistoric times to Kepler and Newton. Partial contents: Near Eastern astronomical systems, Early Greeks, Homocentric Spheres of Eudoxus, Epicycles, Ptolemaic system, medieval cosmology, Copernicus, Kepler, etc. Revised, foreword by W. H. Stahl. New bibliography. xvii + 430pp. 5⅜ x 8.
S79 Paperbound **$1.98**